International Conference on LNG/LPG and Alternative Fuel Ships 2020

London, United Kingdom
29 – 30 January 2020

ISBN: 978-1-7138-0868-8

Printed from e-media with permission by:

Curran Associates, Inc.
57 Morehouse Lane
Red Hook, NY 12571

Some format issues inherent in the e-media version may also appear in this print version.

The Royal Institution of Naval Architects
8-9 Northumberland Street
London, WC2N 5DA
United Kingdom

Phone: 020 7235 4622
Fax: 020 7259 5912

publications@rina.org.uk

Additional copies of this publication are available from:

Curran Associates, Inc.
57 Morehouse Lane
Red Hook, NY 12571 USA
Phone: 845-758-0400
Fax: 845-758-2633
Email: curran@proceedings.com
Web: www.proceedings.com

International Conference on

LNG/LPG and Alternative Fuel Ships

29th-30th January 2020, London, UK

PAPERS

THE ROYAL INSTITUTION OF NAVAL ARCHITECTS
8-9 Northumberland Street, London, WC2N 5DA, Telephone: +44 (0)20 7235 4622

The Royal Institution of
Naval Architects

INTERNATIONAL CONFERENCE

LNG/LPG and Alternative Fuel Ships

29th -30th January 2020, London, UK

CONTENTS

***Unavailable at time of going to press**

LNG BUNKERING VESSELS AND BARGES: A 2020 UPDATE

R.D. Tustin, Lloyd's Register EMEA
J.M. Navarro, **W.J. Choi,** Lloyd's Register Asia

SUMMARY

For ocean going ships where operational endurance, and range, requires larger capacity bunker fuel tanks the adoption of gas as a fuel has established a need for new types of dedicated LNG bunkering vessels (LBV) and LNG bunkering barges (LBB) with specialist systems for handling ship to ship (STS) transfer of LNG fuel.

In 2017 three early examples of purpose designed, and built, LNG bunkering vessels (LBV) entered service to support LNG bunkering operations in European Ports and were described by the authors in a paper [1] presented for publication by RINA in November 2017. In 2018 four more LNG bunkering vessel examples were put into service including a first example of an LNG bunkering barge (LBB) as well as two cases of conversions into LNG bunkering vessels (LBV). These examples were described in a paper [2] presented for publication by RINA-HIMT in December 2018.

Further examples of dedicated LNG bunkering vessels (LBV) and LNG bunkering barges (LBB) have been delivered in 2019.Summarizing the LNG bunkering vessel fleet development the paper will offer an update of the authors 2017 [1] and 2018 [2] papers.

NOMENCLATURE

ATB	Articulated Tug Barge
BASIL	SGMF Bunkering Area Safety Information for LNG model,
BOG	Boil-Off Gas
BOR	Boil-Off Rate
CCS	Cargo Containment System
CEF	Connecting Europe Facility
CPP	Controllable Pitch Propeller
DF	Dual Fuel
DFDE	Dual Fuel Diesel Electric
ECA	Emission control area
ESD	Emergency Shut-Down
EMSA	European Marine Safety Agency
FPP	Fixed Pitch Propeller
GCU	Gas Combustion Unit
GTT	Gas Transport and Technigaz
HFO	Heavy Fuel Oil
IACS	International Association of Classification Societies
IMO	International Maritime Organization
IWW	Inland Waterway
LNG	Liquefied Natural Gas
LOC	Loss of Containment *(event)*
LBB	LNG Bunkering Barge
LBV	LNG Bunkering Vessel
MARPOL	IMO Marine Pollution Convention
NOx	Nitrogen Oxides
NECA	NOx emission control area
SECA	SOx emission control area
SIMOPS	Simultaneous Operations
SGMF	Society for Gas as a Marine Fuel
SOx	Sulphur Dioxide
STS	Ship to Ship
TTS	Truck to Ship
ULCS	Ultra Large Container Ship
VLOC	Very Large Ore Carrier

1. INTRODUCTION

Implementation of stricter limits of emissions of sulphur dioxide (SOx) in Annex VI of the MARPOL Convention, as well as regional SOx emission control areas (SECAs), is encouraging adoption of methane gas, stored as liquefied natural gas (LNG), as a fuel for commercial shipping.

To meet the need for ship to ship (STS) transfer of LNG fuel new types of LNG bunkering vessel (LBV) and LNG bunkering barges (LBB) are now entering service.

1.1 PAPER OBJECTIVE AND SCOPE

The paper objective is to offer a 2020 update on LNG bunkering vessel (LBV) developments adding new projects and insights to reflect fleet development since the 2017 [1] and 2018 [2] papers.

The scope, and content, of this paper includes:

- LNG bunkering vessel (LBV) and LNG bunkering barge (LBB) fleet development including, in this 2020 paper update, ships under construction and on order for delivery in 2021
- An overview of considerations for specification of purpose designed LNG bunkering vessels
- An overview of LNG transfer operations and receiving ship compatibility aspects
- An overview, for this 2020 paper update, of risk assessment and safety zone considerations for simultaneous operations (SIMOPS)

Figure 1: Existing fleet of LNG bunkering vessels, barges and conversion cases

Ship Name	Owner	Type	Year of Delivery	Capacity (m^3)	Conversion case	Port of bunkering operations *(if known)*	1st LNG bunkering operation	
							Date	*Ship name* Ship type
Seagas	Sirius Shipping	LBV	2013	180	✓	Stockholm	03-2013 [3]	*Viking Grace* Ro-Pax
Engie Zeebrugge	Gas4Sea	LBV	2017	5000		Zeebrugge	14-06-17 [4]	*Auto Eco* PCTC
Coralius	Sirius Veder Gas	LBV		5800		-	19-09-17 [5]	*Fure West* Product Tanker
Cardissa	Shell	LBV		6500		Rotterdam	04-10-18 [6]	*Gargarin Prospect* Crude Oil Tanker
Oizmendi	Naviera Urbasa, Ibaizabal Group	LBV	2018	600	✓	Bilbao and Cantabrian Coast	03-01-18 [7]	*Ireland* Cement Carrier
Coral Methane	Anthony Veder	LBV		7500	✓	Santa Cruz de Tenerife	16-01-19 [8]	*AIDAnova* Cruise Ship
Clean Jacksonsville	TOTE Maritime	LBB		2200		Jacksonville	*Not yet published*	*Isla Bella* *Perla Del Caribe* Container Ship
Kairos	Babcock Schulte Energy	LBV		7500		Klaipeda	15-04-19 [9]	*Fure Valö* Product Tanker
LNG London	LNG Shipping	IWW LBB	2019	3000		Rotterdam	06-19 [10]	*Containerships Polar* *Containerships Nord* Container Ship
Flexfueler001	Titan LNG	LBB		1480		Amsterdam	14-06-19 [11]	*Kvitnos* Ro-Ro

2. LBV AND LBB FLEET DEVELOPMENT

At time of writing there are ten (10) LNG bunkering vessels in service comprising:

- Three (3) LNG bunkering vessel conversion cases
- Four (4) purpose designed and built LNG bunkering vessels (LBV)
- Two (2) purpose designed and built LNG bunkering barges (LBB)
- One (1) purpose designed and built Inland Waterway (IWW) LNG bunkering vessels (LBV)

The existing fleet of LNG bunkering vessels is shown in the Figure 1 table.

2.1 LBV AND LBB FLEET ORDERBOOK

At time of writing there are a further twelve (12) LNG bunkering vessels (LBV) under construction and three (3) bunkering barges (LBB) on order, for delivery in 2019, 2020 and 2021 as shown in Figure 2 comprising:

- 2019 deliveries: one (1) purpose designed LNG bunkering vessels (LBV)
- 2020 deliveries: seven (7) purpose designed LNG bunkering vessels (LBV) and two (2) bunkering barges (LBB) (one (1) a purpose designed articulated tug and LNG bunkering barge ATB).
- 2020 conversion: one (1) re-build of a liquid fuel bunker vessel to an LNG bunkering vessel (LBV).
- 2021 deliveries: four (4) purpose designed LNG bunkering vessels (LBV), and one (1) bunkering barges (LBB) (a purpose designed articulated tug and LNG bunkering barge ATB).

Figure 2: LNG bunkering vessels under construction, or on order, for delivery in 2019, 2020 and 2021

Owner	Type	Expected delivery	Capacity (m^3)	LNG Bunker Cargo Tanks			Shipyard
				Type	Construction	Number	
Avenir (formerly Stolt Nielsen Gas)	LBV	12/19	7500	IMO Type C	Cylindrical	2	Keppel Singmarine Nantong Shipyard [12]
Avenir (formerly Stolt Nielsen Gas)	LBV	2020	7500	IMO Type C	Cylindrical	2	Keppel Singmarine Nantong Shipyard [12]
MOL	LBV		18600	Membrane (MkIII Flex)	Prismatic	2	Hudong Zhonghua [13]
Q-LNG	ATB (LBB)		3996	IMO Type C	Cylindrical	4x 999m^3	VT Halter Marine Inc [14]
FueLNG	LBV		7500	IMO Type C	Cylindrical	2	Keppel Singmarine [15]
CLMF Japan	LBV		3500	IMO Type C	Cylindrical	1x 3500m^3	Kawasaki Heavy Ind. (KHI) [16]
Elenger (formerly Eestigas)	LBV		6000	IMO Type C	Cylindrical	2	Damen Yichang Shipyard [17]
ENN	LBV		8000	IMO Type C	Cylindrical	2	Dalian Shipbuilding (DSIC) [18]
Shturman Koshelev	LBV		5800	IMO Type C	Cylindrical	2	Keppel Singmarine [19]
Titan LNG *(Flexfueler002)*	LBB		1480	IMO Type C	Cylindrical	4x 370m^3	Kooiman Shipyard Zwijndrecht [20]
Bergen Tankers	LBV*		Not known	IMO Type C	Cylindrical	1	Note*: re-build of existing liquid bunker tanker into LBV [21]
Polaris New Energy	ATB (LBB)	2021	5400	IMO Type C	Cylindrical	4x1350m^3	Fincantieri Bay Shipbuilding, Sturgeon Bay [22]
Pavilion Energy	LBV		12000	Membrane (MkIII Flex)	Prismatic	2	Sembcorp Marine Tuas Boulevard Shipyard [23]
Avenir	LBV		7500	IMO Type C	Cylindrical	2	Nantong CIMC Sinopacific Offshore & Eng. Co. Ltd. [12]
Avenir	LBV		7500	IMO Type C	Cylindrical	2	Nantong CIMC Sinopacific Offshore & Eng. Co. Ltd. [12]
MOL	LBV		18600	Membrane (MkIII Flex)	Prismatic	2	Hudong Zhonghua [24]

3. LBV SPECIFICATION AND DESIGN

Some basic considerations for specification, and design, of an LNG bunkering vessel (LBV) are summarised in this section.

3.1 LBV ARRANGEMENT AND LAYOUT

LBV designs with cylindrical IMO Type C tanks are arranged with a weathertight non-structural cover of the exposed upper tank dome (s). Platforms may be arranged for cargo manifold, cargo handling equipment and cargo processing rooms on the dome cover.

LBV designs with prismatic membrane cargo containment system are arranged with a continuous strength deck extending over the whole length of ship. Alternatively, a continuous trunk deck may be fitted structurally integrated into a raised poop and forecastle.

Cargo holds are subdivided by transverse bulkhead(s) and, where IMO Type C tanks are installed, partial double skinning may be adopted in way of the cargo holds.

Accommodation and engine room are typically arranged aft. In one case, where Dual Fuel Diesel Electric (DFDE) propulsion system was adopted, the accommodation and engine room has been arranged forward.

For this case the forward accommodation and engine room layout allows for control of trim without the use of sea ballast water. As designed the ship is to be capable of adjustment of trim (and list) by using three pairs of fresh water tanks.

3.2 LNG CARGO TANK TYPE SELECTION

Considerations for selection of LNG bunker cargo tanks for an LNG bunkering vessel (LBV) include:

- Flexibility in partial cargo loading
- Ease of Boil-Off Gas (BOG) management
- Low Boil-Off Rate (BOR)
- Low cost, low weight, maximized cargo volume

Three configurations and two types of LNG cargo tanks have been adopted in the existing fleet of LNG bunkering vessels (LBV) and LNG bunkering barge (LBB) as shown in Figure 3:

- Cylindrical IMO Type C tank
- Bi-lobe IMO Type C tank
- Prismatic membrane type

The main technical advantage of an IMO Type C tank is the ease of pressure build up, or accumulation, as a means of LNG bunker transfer and BOG treatment.

The main technical advantage of a membrane LNG cargo containment system is the possibility of high hull volume utilization and space efficiency.

The main technical drawback of an IMO Type C cylindrical tank is a low hull volume utilization and space efficiency when compared with IMO Type C bi-lobe tank and tanks of prismatic configuration with membrane cargo containment system.

The main technical drawback of an IMO Type C Bi-Lobe is tank weight. For increased LNG bunker volume requirements for receiving ships it is expected that prismatic tanks with membrane cargo containment system will be adopted.

3.3 PROPULSION ARRANGEMENT

Main requirements for propulsion system selection and specification for an LNG bunkering vessel (LBV) are:

- Utilizing LNG (BOG and Vaporized LNG Cargo)
- Redundancy
- Low fuel oil/gas consumption
- High performance in manoeuvrability
- Reasonable initial cost

A single direct drive Dual Fuel (DF) 4 stroke engine together with reduction gear and either fixed pitch (FPP) or controllable pitch (CPP) propeller is expected to be the most economical propulsion system for an LBV. To improve the manoeuvrability, at least one transverse thruster is typically specified.

Alternatives, and additional, specification options for the propulsion arrangements have been adopted for purpose designed LNG Bunkering Vessels in service. These specification options are shown in the Appendix 1, Figure A1 comparison table. Propulsion arrangement options adopted include:

- Twin propellers (FPP or CPP)
- Single/twin rudders with flaps
- Additional transverse bow & stern tunnel thrusters
- Dual Fuel Diesel Electric (DFDE) system
- Azimuthing thrusters (with DFDE)

4. LNG BUNKER TRANSFERS

4.1 METHODS OF LNG BUNKER TRANSFER

LNG bunker transfer operations may be achieved by one of the following methods;

- Ship to Ship (STS)
- Truck to Ship (TTS)
- Terminal to Ship via Pipeline (TPS)

Figure 3: Three configurations and two types of LNG cargo tanks adopted in existing LNG bunkering fleet

IMO Type C tank, cylindrical construction

Image courtesy of STX Offshore & Shipbuilding

IMO Type C tank, bi-lobe construction

Image courtesy of Royal Bodewes

Prismatic construction, Mk III Flex membrane

Image courtesy of Conrad Industries

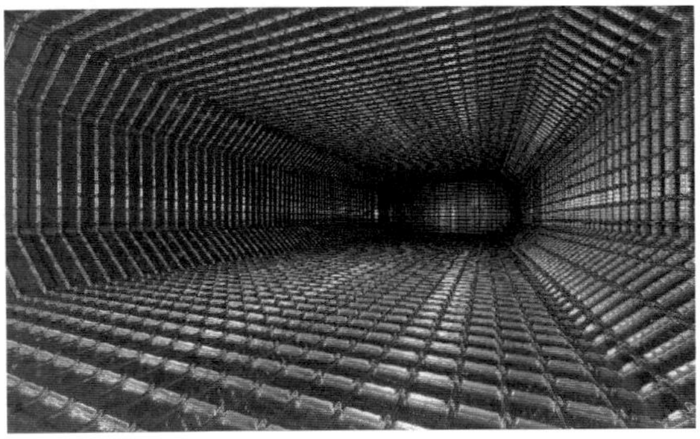

Figure 4: GTT REACH4® *(Refuelling Equipment Arm CH_4)*
LNG bunker loading arm on LBV *Clean Jacksonville*

Image courtesy of Conrad Industries

In practice the method adopted for LNG bunker transfer is dependent on LNG bunkering facilities within the port, the amount of LNG to be transferred and, in the case of Ship to Ship (STS) LNG transfer, the availability of locations for LNG transfer that will ensure uninterrupted, and unimpeded, port operations when the LNG bunkering vessel (LBV) is double banked moored for LNG transfer to a receiving ship.

In a 2013 [25] study the following recommendations for bunkering methods were made based on the bunkering volumes required by receiving vessels:

- Ship to Ship (STS) transfer is recommended for receiving vessels requiring more than 100 m^3 LNG fuel.
- Truck to Ship (TTS) transfer is recommended for receiving vessels requiring up to 200 m^3 LNG fuel.
- LNG Terminal to Ship via Pipeline (TPS) is recommended for all bunkering volumes where there are repeat customers, and receiving vessels, and space available for bunkering facilities

For Ship to Ship (STS) bunkering the study [25] highlighted that one LNG bunker vessel (LBV) per receiving vessel is desirable to minimize port turnaround time. This implies matching of LBV cargo tank size to the bunker tank capacity of the receiving ship. To date LNG bunkering vessels (LBV) have adopted capacities of cargo tanks well matched to that of LNG fuel receiving tanks on gas fuelled ships.

4.2 EQUIPMENT FOR STS LNG TRANSFER

Specification of equipment for Ship to Ship (STS) LNG bunker transfers may include consideration of:

- A loading arm or mast installation, OR
- Use of (cryogenic) flexible hoses

In the existing fleet of ten (10) LNG bunkering vessels there are two (2) examples of loading arm, or loading mast, installations as follows:

- FMC LNG bunkering transfer arm installed on LNG bunkering vessel (LBV) *Cardissa*
- REACH4® LNG loading mast installed on LNG bunkering barge (LBB) *Clean Jacksonville*

Two examples are described below to offer an insight into possible specification considerations for LNG transfer and bunkering interface equipment.

First example is the REACH4® (Refuelling Equipment Arm CH_4) LNG bunker loading mast [26]. Offered by Gas Transport and Technigaz (GTT) this is a rotating and luffing mast of around 20 m boom length. A stated key advantage of REACH4®: a simple emergency disconnect system with breakaway couplings.

The first installation of a REACH4® LNG loading mast is on the 2200m^3 capacity LNG Bunkering Barge (LBB) *Clean Jacksonville* delivered from Conrad Shipyard to TOTE Maritime in August 2018, see Figure 4.

Second example is a hybrid LNG bunker transfer system; combining features of both a loading arm as well as a flexible hose-based transfer.

Developed form a collaboration between KLAW LNG and Houlder Ltd the *KHobra* system is arranged to be deployed using a standard knuckle boom crane [27]. *KHobra* utilizes a customised frame mated to LNG bunker loading and vapour return hoses. All connections are accommodated in a single lift with very significant reductions in connection times which is stated to be a key advantage of *KHobra*.

For controlled rapid shut down of LNG transfer operations during an emergency the LNG bunkering system (as well as the LNG fuel system on-board the receiving vessel) is specified with Emergency Shut-Down (ESD) valves. LNG bunkering vessel (LBV) has also to accommodate different link Emergency Shut-Down (ESD) systems (e.g. Universal Safety Link (USL) system) available.

4.3 MEANS OF LNG TRANSFER

LNG bunker cargo tanks are usually provided with deep well submerged pumps. In accordance with IGC Code, when cargo transfer is by means of non-accessible (e.g. submerged) cargo pumps, two separate means of LNG transfer are to be provided.

The IGC Code accepts the possibility to carry out transfer of the cargo by gas pressurization. Where IMO Type C cargo tanks are selected for an LNG bunkering vessel consideration could be given to the adoption of a primary means of discharge by mechanical pumps and a secondary means via gas pressurization.

To maximize the amount of energy transferred it is desirable for LNG bunker cargo to be carried at its design temperature at atmospheric (1 bar g) pressure. Atmospheric conditions of carriage can also be achieved for IMO Type C cargo tanks and carriage of LNG bunker cargo at atmospheric pressure in an IMO Type C cargo tank allows for pressure accumulation (as one means of cargo transfer) during bunkering operations.

Higher LNG bunker cargo transfer rates generate a higher volume of Boil-Off Gas (BOG) so one possible limitation of LNG bunker cargo transfer rate is the capacity of the LNG bunkering vessel (LBV), as well the LNG bunker receiving vessel, to handle the Boil-Off Gas (BOG).

4.4 MEASUREMENT OF LNG TRANSFER

Custody Transfer System (CTS) is typically in use by LNG carriers to indicate the LNG transfer volumes. In addition, third party laboratory analysis of LNG composition is necessary in a Custody Transfer System (CTS) to measure cargo energy delivery.

For LNG bunker transfers Custody Transfer System (CTS) is not sufficient for spot bunkering cases where there is a need to measure the exact quantity of LNG transfer from LNG bunkering vessel (LBV) to receiving ship and it is recommended to install flow meters at the liquid and vapour return lines. In addition, the composition of the LNG bunker may differ depending of the loading terminal and previously loaded bunkers. It is then also important to control the quality of the LNG being transferred.

For these reasons the use of mass flow meters in combination with calorific value conversion (using a gas chromatograph) into an energy transfer is the preferred option for custody transfer of LNG bunkers.

The Society for Gas as a Marine Fuel (SGMF) has published detailed Guidance for Quality and Quantity of the LNG transfer during bunkering operations [28].

4.5 BOIL-OFF GAS (BOG) HANDLING

Depending on the tank size, and type of insulation, LNG Boil-Off Gas (BOG) may be generated at Boil-Off Rates (BOR) of 0.2% to 0.6% (BOR) of tank volume per day for LNG bunkering vessels (LBV).

Depending on LNG bunker tank arrangements on board the receiving ship may also generate LNG Boil-Off Gas (BOG) at similar Boil-Off Rates (BOR).

However, it is during LNG bunker transfer operations when significant additional Boil-Off Gas (BOG) is generated. This generated Boil-Off Gas (BOG) requires safe handling during the LNG transfer operations without any emissions.

The LNG receiving ship may have its own means to handle the Boil-Off Gas (BOG) generated within LNG bunker fuel tanks.

However, it is also typically expected that the LNG bunkering vessel (LBV) will assist, partially or completely, the receiving vessel to handle the additional Boil-Off Gas (BOG) generated because of LNG bunker transfer operation.

For an LNG bunkering vessel (LBV) a combination of following means will be necessary for Boil-Off Gas (BOG) handling:

- Use of Boil-Off Gas (BOG) as a fuel: for power generation and/ or propulsion.
- Re-liquefaction of Boil-Off Gas (BOG).
- Storage by compression of Boil-Off Gas (BOG): in one LBV case, *Kairos*, BOG is compressed and stored in separate CNG tanks for later use as a fuel
- Pressure build-up, or accumulation, which is feasible in case of IMO Type C LNG tank types
- As a final contingency burning of Boil-Off Gas (BOG) in a Gas Combustion Unit (GCU), or a boiler with dual fuel firing capabilities (where fitted)

A disadvantage of pressure accumulation is that it will cause LNG temperature rise and reduce the transfer rate as pressure increases when vapour return is used during bunkering. The quality of LNG to be delivered will also change as LNG temperature increases, which in turn may lead to higher Boil-Off Gas (BOG) generation during LNG bunker transfer.

Use of Boil-Off Gas (BOG) as a fuel is then the desired, and most economical, approach for Boil-Off Gas Handling, especially as LNG bunkering vessel (LBV) is expected to be operating in an emission and environmentally controlled area.

4.6 LNG RECEIVING SHIP COMPATIBILITY

A critical aspect for the specification and design of an LNG bunkering vessel (LBV) or LNG bunkering barge (LBB) and the associated LNG bunker transfer system is compatibility with expected LNG fuelled receiving ships. To date most LNG fuelled receiving ships have been coupled in long term gas fuel supply contracts associated with nominated ports, and specific LNG bunker vessels and/or barges, and this has facilitated early consideration of LNG bunkering operation compatibility issues.

With increasing numbers of LNG fuelled ships in the world fleet there is a possibility of LNG bunkering operation compatibility issues arising in the case of spot purchases of LNG bunker from incompatible LNG bunkering vessels where LNG fuelled ships are employed in short term worldwide charters.

However, it should be noted that IACS Recommendation 142 [29] and EMSA Bunkering Guidelines [30] require that "A compatibility assessment of the bunkering facility and receiving ship should be undertaken prior to confirming the bunkering operation to identify any aspects that require particular management"

Some potential compatibility arising for LNG bunkering operations are summarized in a Figure 5 table which considers compatibility issues arising from:

- LNG cargo tank and receiving LNG fuel tank types
- Elements of LNG bunker transfer system and LNG fuel receiving system

Figure 5: Summary of potential compatibility issues between LNG bunkering and LNG fuelled receiving ship

LNG bunkering vessel or barge		LNG bunkering operation: compatibility issues	LNG fuelled receiving ship	
LNG cargo tank type	IMO Type C	• Required tank spraying for pressure control and balancing of both tanks, in case of no vapour return • Pressure equalizing through vapour line as an option	IMO Type C	LNG fuel tank type
LNG cargo tank type	IMO Type C	• Spot bunkering case: mismatch of bunker availability and receiving ship fuel demand could be anticipated • The transfer rate adjustment to prevent rapidly pressure increase in the membrane fuel tank should be considered. • Additional equipment (pre-cooler, buffer tank, Pressure Reduction Valves, etc.) will be needed to equalize pressure (between LNG cargo and fuel tank) • The vapour return through vapour line connection is required	Membrane (Atmospheric Tank)	LNG fuel tank type
LNG cargo tank type	Membrane (Atmospheric Tank)	• Transfer rate adjustment to reduce the amounts of return BOG should be considered (BOG Return may not be required) • Flow control valve on the vapour return line will be added to prevent the impact on membrane tank from the amounts of return BOG	IMO Type C	LNG fuel tank type
LNG cargo tank type	Membrane (Atmospheric Tank)	• Vapour return through vapour line connection is required • Transfer rate adjustment and additional BOG management equipment (re-liquefaction, chiller, GCU, CNG buffer tank, etc.) should be considered to handle the amounts of BOG • Alternatively, depending of the capability of the LNG Fuelled ship to handle the BOG generated using 'Hotel Load' (e.g. cruise ships have very high hotel load) additional BOG management equipment may not be necessary	Membrane (Atmospheric Tank)	LNG fuel tank type
LNG bunker transfer system	N_2 Purging	• Nitrogen (N_2) generation (from bunkering or receiving ship) to be decided and required N_2 generation capacity to be considered	N_2 Purging	LNG bunker fuel receiving system
LNG bunker transfer system	Discharge manifold	• Manifold connection compatibility • Elevated manifold to be considered on LBV to account for receiving ship in light draft condition • Mooring arrangement for bunkering operation • Operation radius of hose handling crane incl. fender handling	Receiving manifold	LNG bunker fuel receiving system
LNG bunker transfer system	Bunker loading hose & Vapour return hose	• Develop material handling method of the flexible hose • Emergency Release Connectors (ERC) and Quick Connect/ Disconnect (QCDC) couplings to be considered. • Smallest radius of (hose) curvature considering rolling of two vessels in Ship to Ship (STS) bunker transfer • In case of bunkering to membrane type fuel tank, a vapour return hose is required for safe operation	Bunker loading hose	LNG bunker fuel receiving system
LNG bunker transfer system	Loading arm	• Flexibility to accommodate the connection/disconnection process • Possibility to deliver LNG bunker connection at different heights • Swivel joints allow the necessary motion in the intended degrees of movement	Receiving manifold	LNG bunker fuel receiving system
LNG bunker transfer system	LNG metering	• Mass flow meters + Gas chromatograph is required for a spot bunkering case	LNG metering	LNG bunker fuel receiving system

5. SIMOPS AND SAFETY ZONE STUDIES

For the safe operation of LNG bunkering during the time that the receiving ship is carrying out other operations it is recommended that a SIMOPS (*Simultaneous Operations*) Risk Assessment is performed.

5.1 SIMOPS RISK ASSESSMENT

The purpose of SIMOPS Risk Assessment is to identify Simultaneous Operations (SIMOPs) by the receiving ship that may be completed in parallel with LNG bunkering that could;

- EITHER initiate Loss of Containment (LOC) event
- OR contribute to escalation of LOC event severity

SIMOPS Risk Assessment is to be carried out in accordance with the following ISO standards:

- ISO 31000: 2009, Risk Management – Principles and Guidelines
- ISO 31010: 2010, Risk Management – Risk Assessment Techniques
- ISO 20519: 2017, Ships and marine technology – Specification for bunkering of liquefied natural gas fuelled vessels

The deliverable of a SIMOPS Risk Assessment is a set of recommendations to achieve an inherently safer design through elimination of hazards and prevention of LOC events. Prior to a SIMOPS Risk Assessment study, a safety zone study is performed.

5.2 SAFETY ZONE STUDIES

The primary purpose of a safety zone is to;

- Control ignition sources to reduce the likelihood of igniting a flammable gas cloud due to an accidental release of LNG or natural gas during bunkering
- Support determining an essential minimum number of personnel inside a safety zone exposed to a hazard (e.g. fire) following accidental LNG release

Determination of safety zone distances can be carried out according to methodology indicated ISO/TS 18683:2015 (*deterministic approach*) for LNG ship-to-ship bunkering. ISO/TS 18683 provides detailed guidance for two methodologies to determine safety zones:

- Deterministic methodology based on dispersion analyses for maximum credible release scenario(s)
- Risk-based methodology (QRA) relating to overall safety requirements

The more complex risk-based (QRA) methodology may be used if the deterministic methodology results in too large distances from a practical point of view. The initial hazard identification resulting in loss of containment shall reflect factors specific to the bunkering operation including for example:

- Transfer rates and inventory in bunkering facilities
- LNG temperature and pressure in bunkering system
- Location-specific and representative weather conditions

The maximum effect of the representative scenario shall be used to determine the safety distance. Examples of maximum credible releases are provided in ISO/TS 18683, such as:

- A trapped volume release scenario where LNG in a hose is *'trapped'* between ESD valves OR
- A pressurized release scenario due to a broken connection.

Safety distances are calculated using dispersion modelling of the release rate and duration, considering pressure, environmental conditions and the characteristic dimensions of obstacles in the surroundings. Initially, gas released will be cold and spread at ground level before dispersion of the cloud may occur. Gas may also enter safe areas through ventilation openings and doors.

The most important parameters that could significantly influence the size of the safety zone are the hole size, total released mass, temperature and pressure in the hose, wind speed and stability, surface roughness, availability of liquid spill containment and potential for liquid release to water or land.

The safety zone is implemented as a safety distance all around the bunkering area where originating from the potential LNG release locations. The dimensions are determined by a horizontal safety zone distance for each dispersion surface and a vertical safety zone distance above each dispersion surface.

There is also an SGMF tool, BASIL (*Bunkering Area Safety Information for LNG model*), that can be used to manage bunkering risks through the definition of a safety zone depending on the type of bunkering operation.

5.3 SIMOPS STUDY FINDINGS

In general, it is possible, using a SIMOPS risk assessment to establish that LNG bunkering operations can be carried out concurrently with any other transfers between ship and shore (*or between ships if ship-to-ship bunkering method is used*) including:

- Loading or unloading cargo operations
- Dangerous goods loading or unloading
- Loading or unloading of stores and provisions
- Chemical and other low flash product handling
- Bunkering of fuels other than LNG

6 FUTURE LBV DEVELOPMENT

Future LNG Bunkering Vessel (LBV) specifications and designs will be influenced by developments in LNG tank containment technology, as well as, an expected increased demand for LNG bunker fuel as more commercial ships adopt gas as a fuel.

Described below, as considerations for the specification of a future LNG Bunkering Vessel (LBV), are two possible influences on future LBV designs:

- LNG tank type development
- Adoption of much larger LNG bunker tanks for very large commercial ships

6.1 LNG TANK TYPE DEVELOPMENT

New tank designs and variants of existing LNG tank types are being considered which may be adopted in future LNG Bunkering Vessels (LBV). These include:

- Tri-lobe IMO Type C tank types
- Prismatic IMO Type A or B independent tank types
- Lattice pressure vessel tank types
- Variants of membrane cargo containment systems

The principal driver for these LNG tank developments is improved tank capacities for better hull volume utilization and space efficiency.

The lattice pressure vessel (LPV) is a prismatic pressure containment solution that is scalable covering several combinations of pressure and volume by arranging similar tank design. The tank is provided with internal lattice structure to withstand the internal pressure and assuring structural redundancy. The technical feasibility has been demonstrated using several proto-type tanks. A small size LNG tank, using Lattice Technology, is to be installed [31] as an LNG fuel tank on board a Port Service ship for the port of Ulsan, South Korea.

New design variants of membrane cargo containment systems are to facilitate pressure accumulation as a means of LNG bunker transfer, as well as Boil-Off Gas (BOG) treatment, with an increase in LNG bunker fuel tank pressure above 0.7 bar g.

LNG BRICK is a containment system designed by Gaztransport and Technigaz (GTT) consisting of a prefabricated cubic steel block reinforced to support a design pressure up to 2.2 bar g and having the steel structure thermally insulated using a GTT Mark III membrane technology. BRICK steel structure block is intended to be integrated into the vessel structure during ship construction. The tank design can accommodate a LNG Capacity up to 3,000 m3. The tank pressure and its insulation is designed to contain the Boil Off Gas (BOG) for 15 days. In March 2018, a construction of a prototype LNG BRICK tank was completed [32].

6.2 LARGER LNG BUNKER TANKS

Ultra large container ships (ULCS) of 20,000 teu cargo capacity may consume about 500m^3 of LNG per day. If endurance for a round-trip on an Asia-Europe service is required this would correspond to 40 days operations and require LNG bunker capacity of about 22,000m^3.

If Ship to Ship (STS) LNG bunker transfer is anticipated for a gas fuelled Ultra Large Container Ship (ULCS) the following conditions for the LNG bunkering operation could be anticipated:

- A short port turnaround time, typically, in duration between 12 to a maximum of 36 hours.
- A possible large size of LNG fuel tank depending on the expected LNG fuelled Ultra Large Container Ship (ULCS) endurance.
- A significant preparation time, e.g. time for cool down of receiving fuel system, for the LNG fuelled Ultra Large Container Ship (ULCS) prior to LNG commencing bunkering operations.
- A possible requirement for a high volume of LNG bunker fuel transfer, in case of round-trip endurance, for LNG fuelled Ultra Large Container Ship (ULCS).

To meet these anticipated, and demanding, conditions for bunkering of LNG fuelled Ultra Large Container Ship (ULCS) the following may need be considered for LNG bunkering vessel (LBV) specification and design:

- A possible increase in LNG bunker cargo transfer rate, in proportion to the increase in volume of LNG bunker to be transferred, to the LNG fuelled receiving ship.
- Faster connection, and disconnection, speed for LNG bunkering interfaces encouraging the adoption, and installation, of specialised and purpose designed LNG bunker transfer equipment.
- An optimized bunkering process to enable the receiving LNG fuelled ship to load LNG within the short available port turnaround time.

Since the publication of our original paper in November 2017 the first orders for a 18600m^3 capacity LNG bunkering vessel have materialised for supply of LNG fuel to a fleet of gas fuelled ultra large container ships (ULCS) [33].

In addition, there are other large commercial ships, such as very large ore carriers (VLOC) of 350,000 to 400,000 dwt tonne size for trade from Brazil to China, where hull space has been allocated for future installation of large LNG bunker tanks >25,000m^3 capacity for round-trip endurance (on conversion to gas as a fuel operation).

7. SOME ALTERNATIVE LBV SOLUTIONS

Some alternatives, to dedicated and purpose designed LNG bunkering vessels (LBV) and bunkering by Ship to Ship (STS) LNG bunker transfer, described below include:

- Multi-fuel bunkering vessels
- Open cargo deck possibilities for LNG bunker cargo tanks and bunker supply system
- Dual use ship with plug in LNG tanks for bunkering operations

7.1 MULTI-PURPOSE BUNKERING VESSELS

The conversion of liquid-fuel bunkering vessels to multi-fuel bunkering vessels (i.e. fuel oil, gas oil and LNG) by the installation of IMO Type C LNG deck tanks, either at the new-building stage or as a conversion, has been considered especially for Atlantic and Mediterranean ports in Spain.

The first conversion of an oil recovery and bunker fuel oil vessel *Monte Arucas* to the multi-fuel bunkering vessel *Ozimendi* was delivered in 2018.

Such a solution is advantageous for flexibility of bunkering vessel deployment in the current situation where the numbers of gas fuelled ships in the world fleet is comparatively small.

However, for conversion cases the installation of deck tanks may imply a reduction in the liquid fuel carrying capacity of a converted vessel, or in the case of a newbuilding, a larger vessel to achieve a desired deadweight of liquid fuel cargo (with added lightweight of LNG deck cargo tanks and LNG bunker fuel transfer system elements).

7.2 OPEN CARGO DECK POSSIBILITIES

In the case of the *Seagas* conversion an available open ro-ro cargo deck was found to be suitable for installation of ro-ro deck mounted IMO Type C LNG bunker cargo tank and LNG bunker fuel transfer system.

There are other ship types with open cargo deck areas that could also be suitable for conversion to an LNG bunkering vessel (LBV) by installation of deck mounted IMO Type C LNG bunker cargo tanks and bunkering equipment. One candidate ship type for such a conversion could, for example, be an offshore supply vessel utilizing the stern cargo deck area for LNG bunker cargo tank and bunkering equipment.

7.3 PLUG IN LNG TANKS FOR BUNKERING

A first case of adoption, for a commercial Ro-Ro vessel, of trailer mounted LNG tanks and connection to a deck mounted fuel supply system has been implemented.

In this first example case three 51.0 m^3 capacity, 7.0 bar g pressure, LNG tanks mounted on road trailers are roll-on loaded at Melbourne for round trip voyage to Tasmania (Bass Strait). This example Ro-Ro ship, *Searoad Mersey II,* entered service in 2017 [34].

Such a scheme for plug in of trailer mounted, or alternatively, ISO container frame mounted LNG tanks may possibly be a suitable solution to consider for a plug in, dual-use adaptation possibility of existing small Ro-Ro and Container ships as LNG bunkering vessels.

One drawback of such an arrangement, and similarly for ISO container frame mounted LNG containers, is kerbside weight limitations for landside roads of the truck/trailer combined unit which limits tank weight and LNG tank capacity.

8. CONCLUDING REMARKS

At time of writing there are ten (10) LNG bunkering vessels in service comprising:

- Three (3) LNG bunkering vessel conversion cases
- Four (4) purpose designed and built LNG bunkering vessels (LBV)
- Two (2) purpose designed and built LNG bunkering barges (LBB)
- One (1) purpose designed and built Inland Waterway (IWW) LNG bunkering vessel (LBV)

Three configurations, and two types, of LNG cargo tanks have been adopted for LNG bunkering vessels (LBV) and LNG bunkering barges (LBB) in service, under construction and on order;

- Cylindrical IMO Type C tank
- Bi-lobe IMO Type C tank
- Prismatic membrane type

At time of writing there are a further twelve (12) LNG bunkering vessels (LBV) and three (3) bunkering barges under construction, or on order, for delivery in 2019, 2020 and 2021.

New projects are being contracted with increasing frequency and the consideration of fleet development, as well as the specification and design, of LNG bunkering vessels and barges is a subject that the authors plan to re-visit with future paper updates.

9. ACKNOWLEDGEMENTS

The authors wish to thank Lloyd's Register EMEA and Asia for supporting paper publication. The authors wish to also thank industry contributors for consent to use credited photographs and reference documents. Any views expressed are from the authors alone and do not necessarily represent those of Lloyd's Register.

10. REFERENCES

1. RINA, 'Purpose designed LNG bunkering vessels (LBV): some characteristics and features', *Conference on Design, Construction & Operation of LNG/LPG Ships,* November 2017
2. RINA-HIMT, 'LNG bunkering vessels and barges: a 2018 update', *Conference on Design, Construction & Operation of LNG/LPG Ships,* December 2018
3. PRESS RELEASE, 'First port in the world to provide an infrastructure for liquefied natural gas for large passenger ferries', *Ports of Stockholm*, 22nd March 2013
4. JOINT PRESS RELEASE (ENGIE, MITSUBISHI CORPORATION, NYK, FLUXYS, PORT OF ZEEBRUGGE, UECC), 'Ship-to-ship LNG bunkering service started in the port of Zeebrugge', *Port of Zeebrugge,* 14th June 2017
5. TRADEWINDS NEWS ARTICLE, 'SkanGas marks first Ship-to-Ship LNG bunkering', *Tradewinds*, 19th September 2017
6. PRESS RELEASE, 'Shell and Sovcomflot LNG bunkering in Rotterdam marks number of firsts', *Port of Rotterdam*, 4th October 2018
7. PRESS RELEASE, 'Oizmendi completes first pilot LNG bunkering (ship-to-ship loading) operation …', *Bilbao Port*, 3rd February 2018
8. PRESS RELEASE, 'Coral Methane successfully executes LNG bunkering operation on AIDANova', *Anthony Veder*, 17th January 2019
9. LNG WORLD NEWS ARTICLE, 'Nauticor's Kairos supplies LNG to Furetank's Fure Valö', *LNG World News*, 16th April 2019
10. NEWS & MEDIA RELEASE, 'LNG London bunker vessel begins operations in Europe', *Shell Global*, 24th June 2019
11. NEWS RELEASE, 'Titan LNG's Flexfueler delivered first LNG', *Titan LNG*, 17th June 2019
12. LNG WORLD NEWS ARTICLE, 'Avenir LNG switches order for LNG carrier duo from Keppel to SOE', *LNG World News*, 15th April 2019
13. PRESS RELEASE, 'Total and MOL sign a long-term charter contract for a pioneer Liquefied Natural Gas (LNG) bunker vessel', *Total S.A.*, 6th February 2018
14. PRESS RELEASE, 'VT Halter Marine starts construction on America's first offshore LNG ATB unit', *VT Halter Marine*, 8th March 2018
15. PRESS RELEASE, 'FueLNG orders South East Asia's first LNG bunkering vessel', *Keppel Corporation,* 5th June 2018
16. PRESS RELEASE, 'Kawasaki Receives First Order for Domestic-use LNG Bunkering Vessel', *Kawasaki*, 6th July 2018
17. PRESS RELEASE, 'Eesti Gaas places order for the first LNG bunker vessel for North-East Baltic Sea', *Eesti Gaas*, 2nd October 2018
18. LNG WORLD NEWS ARTICLE, 'ENN orders LNG bunkering vessel at DSIC', *LNG World News*, 19th October 2018
19. LNG WORLD NEWS ARTICLE, 'Keppel scores ice-class LNG bunkering vessel build', *LNG World News*, 26th December 2018
20. NEWS RELEASE, 'Fluxys and Titan LNG to build LNG bunkering pontoon for the Antwerp port and region', *Titan LNG*, 5th February 2019
21. LNG WORLD NEWS ARTICLE, 'Høglund to provide gas solutions to Norway's first LNG bunker vessel', *LNG World News*, 28th August 2018
22. NEWS RELEASE, 'Northstar midstream to form LNG marine transportation company', *JAXLNG*, 23rd August 2019
23. PRESS RELEASE, 'Pavilion Energy charters its first LNG bunker vessel newbuild for Singapore', *Pavilion Energy*, 25th February 2019
24. SHIP TECHNOLOGY NEWS, 'Total and Mitsui OSK Lines sign agreement for LNG bunker vessel', *Ship Technology*, 5th December 2019
25. DANISH MARITIME AUTHORITY, 'North European LNG Infrastructure Project: A feasibility study for an LNG filling station infrastructure', *Danish Maritime Authority*, 2012
26. LNG INDUSTRY MAGAZINE ARTICLE, 'Unlocking the LNG Fuel Supply Chain', *LNG Industry magazine,* October 2015
27. HOULDER/KLAW LNG PRESENTATION, 'The Houlder-KLAW "Khobra" Transfer System', *Houlder Ltd,* April 2017
28. THE SOCIETY FOR GAS AS A MARINE FUEL, 'Contractual Guidelines Quantity and Quality', *Society for Gas as a Marine Fuel*, September 2015
29. IACS RECOMMENDATION No. 142, 'LNG bunkering guidelines', *IACS*, June 2016
30. EMSA, 'Guidance on LNG Bunkering to Port Authorities and Administrations', *EMSA*, 31st January 2018
31. PRESS RELEASE, 'LATTICE Technology secures first commercial sale of its on-board LPV LNG Fuel tank', *LATTICE Technology*, 24th May 2018
32. GTT PRESENTATION, 'LNG BRICK ceremony', *GTT (Gaztransport & Technigaz)*, March 2018
33. PRESS RELEASE, 'CMA CGM is the first shipping company to choose liquefied natural gas for its biggest ships', *CMA CGM,* 7th November 2017

34. MAN CYRO PRESENTATION, 'Case Study – LNG trailer fuel gas system – Searoad Mersey II (Slides 25 to 34)', *MAN Cyro*, 5th June 2018
35. ROYAL INSTITUTION OF NAVAL ARCHITECTS, 'Significant Ships of 2016 - ENGIE ZEEBRUGGE: LNG bunkering vessel', *RINA*, 2017
36. TGE MARINE GAS ENGINEERING DATASHEET 2735, '5,100 cbm LNG Bunker-Vessel *"Engie Zeebrugge"* for Gas4Sea (Engie, NYK, Mitsubishi)'', *TGE Marine Gas Engineering GmbH*, 2017
37. ROYAL BODEWES WEBSITE DATASHEET, '5800 m^3 LNG carrier', *Royal Bodewes*, 2017
38. TGE MARINE GAS ENGINEERING DATASHEET 2826, '5,800 cbm LNG Bunker Vessel *"Coralius"* for Sirius Veder Gas AB', *TGE Marine Gas Engineering GmbH*, 2017
39. STX OFFSHORE & SHIPBUILDING POCKET PLAN, 'Shell 6.5 K LNG Bunkering Vessel', *STX O & S*, 2014
40. TGE MARINE GAS ENGINEERING DATASHEET 2960, '6,500 cbm LNG Bunker-Vessel *"Cardissa"* for Shell', *TGE Marine Gas Engineering GmbH*, 2017
41. HYUNDAI MIPO DOCKYARD POCKET PLAN, '7,500 m3 Class LNG bunkering vessel', *Hyundai Mipo Dockyard*, 2017
42. LNG SHIPPING 2017 PRESENTATION, 'LNG gas supply vessels – setting the standard', *Babcock LGE Process*, June 2017
43. GTT PRESENTATION, 'Gas ship and gas fuel technologies', *GTT (Gaztransport & Technigaz)*, 9th June 2017
44. TITAN LNG PRESENTATION, 'Flexfueler: the low-cost solution to enable LNG bunkering', *Titan LNG*, 24th September 2018
45. PRESS RELEASE, 'VT Halter Marine signs LNG bunkering ATB with Q-LNG', *VT Halter Marine*, 14th November 2017
46. GA DRAWING, '2271-2050-01 Rev 4, 4,000 m^3 LNG ATB Barge General Arrangement', *VT Halter Marine*, 25th October 2017

11. AUTHORS BIOGRAPHY

Woo-Jeong Choi holds the current position of Lead Sales/Client Relationship Manager at Lloyd's Register Asia. Prior to joining Lloyd's Register, he has had 24 years of shipyard experience includes sales engineering, shipbuilding control, commissioning, trials and project management involvement in projects for the build of LNGC and LBV.

Jose Navarro holds the current position of Global Gas Technology Specialist with Lloyd's Register. He provides expert advice to Clients globally on gas technology and contributing to the creation of specialist rule and regulatory requirements. He has over 20 years' experience of gas technology and newbuilding in Korea, Japan and China.

Robert Tustin holds the current position of Technical Consultant, New Construction for Lloyd's Register EMEA. He is responsible for delivery of pre-tender and pre-contract consultancy advice for non-standard and special ship types. He has 22 years' experience with ship new construction based in North Asia.

APPENDIX 1: LBV AND LBB DESIGNS COMPARED

Figure A1: Comparison of purpose designed LNG bunkering vessel (LBV) in service

Ship name	***ENGIE Zeebrugge***	***Coralius***	***Cardissa***	***Kairos***
Cargo capacity	5100 m^3	5800 m^3	6500 m^3	7500 m^3
Designer	Hanjin Heavy Industries, Busan, Republic of Korea	FKAB, Uddevalla, Sweden	STX Offshore & Shipbuilding, Jinhae, Republic of Korea	Hyundai Mipo Dockyard, Ulsan, Republic of Korea
Shipbuilder		Royal Bodewes*, Hoogezand, The Netherlands		
Yard/NB/Hull No.	268	803	4080	8250
Classification	Bureau Veritas	Bureau Veritas	Lloyd's Register	Lloyd's Register
Main Dimensions				
LOA	107.6 m	99.7 m	119.9 m	117.1 m
LBP	101.8 m	92.2 m	113.1 m	113.5 m
Breadth mld, B	18.4 m	18.0 m	19.4 m	20.0 m
Depth mld, D	9.0 m	12.1 m	10.9 m	10.3 m
Draft design, Td	4.7 m	5.7 m	5.8 m	5.2 m
Deadweight and Tonnage				
Deadweight	2990 tonnes	3100 tonnes	4498 tonnes	4360 tonnes
Gross tonnage	7403	6105	9816	8070
LNG Cargo Tanks				
Cargo tank designer	TGE-Marine	TGE-Marine	TGE-Marine	Babcock LGE
Type	IMO Type C	IMO Type C	IMO Type C	IMO Type C
Construction	Cylindrical	Bi-Lobe	Cylindrical	Cylindrical
Number of tanks	2x 2550 m^3	1x 2550 m^3 1x 3250 m^3	2x 3250 m^3	2x 3750 m^3
Tank material	9% Ni Steel	9% Ni Steel	9% Ni Steel	9% Ni Steel
Design pressure	4.0 bar g	4.5 bar g	3.5 bar g	3.75 bar g
LNG Bunker Transfer				
Bunkering rate	600 m^3/h	840 m^3/h	1,100 m^3/h	1,150 m^3/h
Hose handling crane	2x telescopic type	1x single boom type 1x knuckle boom type	1x single boom type (@elevated manifold)	2x knuckle boom type
Transfer arm	-	-	1x FMC transfer arm	-
Boil-Off Gas Handling				
Boil-Off as fuel	Y: Main Engine Aux. Engines Boiler	Y: Main Engine Aux. Engines Boiler	Y: DFDE Boiler -	Y: DFDE - -
Boil-Off burning	Y: GCU	Y: GCU	Y: GCU	N: -
Re-liquefaction	N: -	N: -	Y: 8.2 tonnes/day	N: -
Compression	N: -	N: -	N: -	Y: 220 bar CNG
Pressure build-up	Y: $\leq$ 4.0 bar g	Y: $\leq$ 4.5 bar g	Y: $\leq$ 3.5 bar g	Y: $\leq$ 3.75 bar g
Propulsion System				
Propulsion	DF direct drive	DF direct drive	DFDE	DFDE
Main engines – Type	2x Wartsila 9L20DF	1x Wartsila 6L34DF	3x Wartsila 8L20DF	3x Wartsila 8L20DF
Main engines – Power	3,330 kW (2x 1,665 kW)	3000 kW	4,260 kW (3x 1,420 kW)	4,800 kW (3x 1,600 kW)
Propeller type	2x CPP	1x CPP	2x FPP	2x azimuth thrusters (2x 1,900 kW)
Rudder type	2x spade with flap	1x spade	2x spade with flap	
Speed (knots)	13.1	13.5	13.0	12.5
Thruster(s) – Bow	2 x 270 kW	1x 600 kW (Retractable)	1x 650kW	2x 320kW
Thruster(s) – Stern	-	1x 400 kW	-	-
Source references	[35] [36]	[37] [38]	[39] [40]	[41] [42]

Note* hull built by Gdanska Stocznia *Remontowa*, Gdansk, Poland

Figure A2: Comparison of purpose designed LNG bunkering barges (LBB) in service and under construction

Ship name	***Clean Jacksonville***	***FlexFueler001***	Not yet published: Owner Q-LNG
Cargo capacity	2200 m^3	1480 m^3	3996 m^3
Designer	Bristol Harbour Group	Kooiman Shipyard Zwijndrecht	Waller Marine
Shipbuilder	Conrad Orange Shipyard		VT Halter Marine
Yard/NB/Hull No.	H489	*Not known*	2030
Classification	American Bureau of Shipping	Bureau Veritas	American Bureau of Shipping
Delivery date	08-2018	06-2019	1Q2020 *Expected*

Main Dimensions

			Barge	Tug	ATB
LOA	70.7 m	76.5 m	98.8 m	39.0 m	125.8 m
LBP	-	-	96.6 m	-	121.3 m
Breadth mld, B	14.9 m	11.4 m	19.5 m	12.8 m	19.5 m
Depth mld, D	4.9 m	3.8 m	9.9 m	6.4 m	9.9/6.4 m
Draft design, Td	2.5 m	-	3.66 m	5.5 m	5.5 m

LNG Cargo Tanks

Cargo tank designer	GTT *(Gaztransport & Technigaz)*	Cryovat International	Wartsila
Type	Membrane	IMO Type C	IMO Type C
Construction	Prismatic	Cylindrical	Cylindrical
Number of tanks	1x 2200m^3	4x 370m^3	4x 999m^3
Tank material(s)	Mk III Flex patented type	*Not known*	*Not known*
Design pressure	0.7 bar g	4.0 bar g	*Not known*

LNG Bunker Transfer

Bunkering rate	500 m^3/h	30 m3/h to 600 m^3/h	1,100 m^3/h
Hose handling crane	-	1x single boom type	2x single boom type
Transfer arm	1x 20m REACH$_4$ transfer mast *(Refuelling Equipment Arm CH_4)*	-	-

Boil-Off Gas Handling

Boil-Off as fuel	N	-	N	Truck loading of LNG on demand		*Not known*
Boil-Off burning	N	-	N	No active Boil-Off Gas management	N	-
Re-liquefaction	Y	6x StirLNG-4 type ~ 6 tonnes/day	N		Y	Air Liquide Turbo-Brayton (TB 350)
Compression	N	-	N		N	-
Pressure build-up	Y	≤ 0.7 bar g	Y	≤ 4.0 bar g	Y	≤ tank design pressure

Barge Power Generation System

Gas or HFO		*Not known*	*Not known*	Gas*
Generators	Type	*Not known*	*Not known*	*Not known*
	Power	*Not known*	*Not known*	2250 kW
Thruster(s)	Bow	-	*1x not known*	*1x not known*
	Stern	-	-	-
Source references		[43]	[44]	[45] [46]

Note* tug in ATB combination NOT gas fuelled

CONVERSION OF AN LNG CARRIER TO FLOATING STORAGE AND REGASIFICATION UNIT

R. Bhat, University of Strathclyde, UK

SUMMARY

Natural gas is hydrogen rich mixture of light hydrocarbons having the chief composition of methane. While natural gas can be transported through pipelines, the cost of pipeline is between 3- 4 million USD per km (Songhurst et al., 2017) which makes the transportation of natural gas over large distances cost ineffective and hence rest of the natural gas trade is by LNG through LNG carriers. LNG trade volume will increase to 50 % from 2014 to 2020 (Shell-LNG-outlook, 2017). Since the increase in natural gas demand will be sudden, many countries will are required to augment their LNG import infrastructure including regasification units, Storage units like tanks. While the development of import infrastructure can take more money and time, like up to 4 years. A new build FSRU can take up to 36 months and require CAPEX of up to approx 450 million USD (Egashira et al., 2013). This cost of FSRU can be brought down by converting an LNG carrier costing approx 320 million USD and the time of 12-24 months (Songhurst et al., 2017). This project was on conversion of an LNG carrier to FSRU to provide 2.5 MTPA regasification facility in Jaigarh port, India.

NOMENCLATURE

[Symbol]	[Definition] [(Unit)]
P	power (kW)
q	Flow capacity (m^3/h)
q_m	Flow capacity (ton/h)
ρ	density of fluid (kg/m^3)
g	acceleration due to gravity (9.81 m/s^2)
h	differential head (m)
c	specific enthalpy of natural gas (kJ/ton)
s	specific heat of sea water (kJ/ton C)

1. INTRODUCTION

Natural gas is hydrogen rich mixture of light hydrocarbons having the chief composition of methane and is very environmental friendly and hence is called a less emission intensive fossil fuel. While natural gas can be transported through pipelines, the cost of pipeline is between 3- 4 million USD per km (Songhurst et al., 2017) which makes the transportation of natural gas over large distances cost ineffective and hence rest of the natural gas trade is by LNG through LNG carriers. Realizing this many countries are preparing to increase the proportion of natural gas in their energy mix after COP-21 summit held in Paris. The new countries expected to have greater share in import of natural gas will be developing countries like China, India, Pakistan and Jamaica. The global LNG trade was 258 million tons in 2016 with an increase of 13 million tons compared to 2015 when there was a record of 244.7 million tons over the LNG trade of 241.3 million tons in 2014. LNG trade volume will increase to 50 % from 2014 to 2020 (Shell-LNG-outlook, 2017). Since the increase in natural gas demand will be sudden, many countries will are required to augment their LNG import infrastructure including regasification units, Storage units like tanks. While the development of import infrastructure can take more money and time, like up to 4 years. The countries with access to sea can utilize the facilities like FSRU (Floating storage and regasification unit), FSU(Floating storage unit) to leverage on the current LNG market with long-term and short term flexibility due to spread of natural gas fields across the globe (World LNG report, 2017). It should also be noted that a new onshore regasification terminal requires a total CAPEX of 750 million USD and the construction requires a time of up to 48 months, whereas a new build FSRU gets completed in up to 36 months and requires a total CAPEX of 450 million USD (Egashira et al., 2013). This cost of FSRU terminal can further be brought down by carrying out conversion of an LNG carrier, which makes the cost up to 320 million USD and the time within 24 months (Songhurst et al., 2017). It is to be noted that the time of 24 months for conversion of LNG carrier includes the time of ordering and procurement of process equipment which can be brought down by making certain arrangements like the one by Höegh LNG which has signed agreements with MOSS Maritime and Wärtsila Oil and Gas (EPCIC regas) for its FSRU conversion project which allows the ordering of critical equipment with long lead times within 12 months (David et al., 2016) and hence using such arrangements the lead time of a FSRU conversion project can be brought down to 12 months. Due to the come up of new natural gas reservoirs across the globe like US and Pacific basin, there has come a widespread distribution of liquefaction capacities across the globe thus reducing the spot charter rates and making LNG more readily and flexibly available, this has further reduced the cross basin trade and only new and large LNG carriers are finding market place, which provides perfect opportunity for the conversion of existing LNG carriers to LNG FPSOs including FSRUs. After COP 21 summit, India is determined to reduce its dependence on fossil fuels and switch to environment friendly fuels like renewables and natural gas. India's production of natural gas at the end of 2016 was 27.6 tons compared to its consumption of 50.1 tons, due to which it imported 22.5 tons from countries like USA, Brazil, Peru, Trinidad, Norway, Qatar and Nigeria. While the import of 22.5 tons is supported by the current LNG import infrastructure of India, after India's resolution to switch to gas fuel will create a demand of more 60 million tons in 2020 (Sen et al., 2017) and it is required to have an import infrastructure for India to be readily available. FSRU or A floating storage and regasification unit is an LNG terminal whose main structure is a special ship that is moored next to the port and can provide a cost effective LNG import infrastructure. While On shore terminals provide more permanent solution, FSRUs allow greater flexibility by allowing operations in offshore and solve problems like space constraints, also requiring less CAPEX costs. There was a proposal of 2.5 MTPA regasification terminal in Jaigarh port in the Maharashtra state of India. While an onshore terminal project will take upto 4 years and hence cannot satisfy the demand at present, which India requires currently after the COP 21 summit, an FSRU can be provided as an early gas option before the completion of the onshore terminal project of Jaigarh port. Like in Brazil Pecem FSRU terminal was closed in 06-2017 before being replaced by permanent onshore terminal as FSRU providers can replace vessels according to their needs. In this project an endeavour has been made to convert a 152,880 m3 moss LNG carrier to FSRU. Design of this project is composed of the conversion concept, the marine design and the process system design. Conversion concept demonstrates how the LNG carrier will be converted to FSRU, the modifications that will be required and the processing systems that will be required along with the layout. The marine design pertains to naval architecture aspects like vessel's hull/size, LNG containment type, mooring, unloading system, etc., while the process design focuses on the process system employed for FSRU. The designs and the technology employed in carrying out the conversion to FSRU were based on project's design criteria. Since, in this project, additional systems will be installed in the LNG carrier, this will increase the power requirements of the FSRU. Cost analysis and safety assessment of the project is done justify the economic and safety worthiness of the project. The cost analysis provides useful and realistic data along with analysis for the comparison of the project with FSRU new builds, and onshore regasification infrastructures along with the justification of the business and economic worthiness of the project.

2. SCOPE AND OBJECTIVES

The scope of this project is to put forth a novel conversion concept to convert an LNG carrier to FSRU to improve/ optimize on economic and safety worthiness of a conversion or FSRU facility. The wider impact of this project is:

- To spark new investments in natural gas import infrastructure.
- To achieve faster realization of projects with quick FIDs (Final Investment decision)
- To provide a cost effective early gas import infrastructure before the decision of building an onshore terminal.
- To cover energy deficit by providing cost effective utilization of low cost LNG.
- To achieve redeployment of laid-up LNG carriers/vessels.

Project's measurable objectives are,

- To provide an economically prospective and cost effective natural gas infrastructure for natural gas consumer countries.
- To provide natural gas infrastructure project that can be implemented in a shorter schedule and can be provided readily as compared to new builds, onshore terminals and other available import alternatives.

3. METHODOLOGY

- Investigation of Natural gas import infrastructure in India.

India's production of natural gas at the end of 2016 was 27.6 tons compared to its consumption of 50.1 tons, due to which it imported 22.5 tons from countries like USA, Brazil, Peru, Trinidad, Norway, Qatar and Nigeria. While the import of 22.5 tons is supported by the current LNG import infrastructure of India, after India's resolution to switch to gas fuel will create a demand of more than 50 million tons in coming years. For this there is a proposal of LNG terminals in Gangavaram port for 5 MMTPA, Kolkata Port for 2.5 MMTPA, Jaigarh Port for 2.5 MMTPA, Chhara Port for 5.0 MTPA, Krishnapatnam for 2.5 MMTPA.

- Selection of the port for the project.

For the project, Jaigarh terminal is selected to provide 2.5 MMTPA regasification facility.

- Evaluation of design criteria, met ocean conditions including average wave height, wind speed.
- Preparing marine design and process system design of the FSRU.

Marine design provides information on the naval architecture aspects of the FSRU including the hull specifications like size, carriage capacity, LNG storage capacity, draft, Cb etc.; this should be complemented by information on LNG containment system, unloading system, project life and mooring system. Process System design provides information like send out rate of natural and the equipment specifications and layout of the process equipment in the converted FSRU.

- Carrying out investigation of the LNG fleet.
- Making selection of LNG carrier/vessel based on design criteria, marine Design requirements of the FSRU.

- Making out the ship modifications required to make the conversion to FSRU.
- Process system design and layout.
- Conducting cost analysis.

4. DESIGN

4.1 DESIGN CRITERIA

Design criteria is the main driver for any project related to an offshore LNG facility like FSRU and includes details like port specifications and met ocean conditions in the area of FSRU operation.

4.1 (a) PORT SPECIFICATIONS

JSW Jaigarh Port is a deep water 24/7 private port in the Maharashtra state of India. It is a Greenfield port and occupies a strategic location in the west coast of India between Mumbai and Goa. Geographical coordinates of the port are latitude 17°18'09" N and longitude 73°12'30" E.
The port can accommodate a vessel size of LOA 295 m, Breadth of 42m.
The port provides a breakwater length of 712 m and draught of 18.5 m.

4.1 (b) MET OCEAN CONDITIONS

Met ocean is the marine analogy of topography and is the abbreviation of meteorology and oceanography. Analysis of met ocean conditions include analysing wind speed, height of wave, temperatures of water and air, speed of the currents and direction and impact of tides etc.
The met ocean statistics of the port were obtained from hindcast in Metoceanview website. The met ocean conditions in this area are significant, though the port infrastructure provides adequate relief from them, but in the design of this project, the port infrastructure like breakwater has not been considered. The average wind speed reaches 15 m/s (1hr.) and the wave height ranges between 6-8 meters. However, the maximum wave height can occasionally reach 8 meters and the maximum wind speed 27 m/s. There is no occurrence of seasonal and occasional icebergs in Indian waters. Met Ocean conditions data is used to select the right offloading systems and while designing the top side structure of FSRU. Since in this conversion the process system is inside the moss tank, such analysis is not required since the stability of FSRU in this case is similar to the original LNG carrier.

4.2 MARINE DESIGN

4.2 (a) PROJECT LIFE, STORAGE CAPACITY AND UNLOADING PERIOD

The project is on the conversion of an existing LNG carrier to Floating station and regasification unit. Since the design life of any LNG carrier is generally 20 years as per DNV, ABS and other classification codes, in this project we carry out conversion of a 7 year old LNG carrier, hence the project life of the project is taken to be 10 years.

The FSRU will regasify 2.5 MMTPA (Million tons per annum) of Liquefied natural gas.
The LNG storage capacity of the FSRU will be 114,600 m3 which corresponds to 52,456 tons (Density of LNG = 0.4577 g/cm3 = 0.4577 ton/m3)
Hence this storage needs to filled 47 times in one year (2.5 MMT/ 52,456 ton = 47)

Corresponding to the requirement of 47 unloadings of LNG in one year we get the unloading period of 7 days (350 days/ 47 unloadings= 7 days/ unloading)
Now it required to provide contingency period, which in this project is chosen to be 2 days
Hence the first unloaded quantity of LNG will be 114,660 m3 and after this every 5 days, the unloaded quantity will be 76,440 m3 and the remaining 38,220 m3 can be used as a contingency for 2 days in case of need.

4.2(b) LNG CONTAINMENT SYSTEM

Selection of LNG (Liquefied natural gas) Containment System for FSRU depends on the following factors:

- Met ocean conditions in the area of operation of FSRU.
- Foot Print requirements for the accommodation of the Process System.
- Storage requirements

Met ocean conditions like waves and winds induce ship motions like sway, surge, heave, yaw, pitch and roll motions (6 degrees of freedom), which result in the fluids motions (In 6 degrees of freedom) inside the containment system which result in fluid waves. These fluid waves can damage the containment system in case of the partially filled tank, which is commonly known as "Sloshing effect". Sloshing can lead to resonance if excitation frequency approaches the natural frequency of the free surface of LNG (Yu et al., 2017) and cause significant damage to the tank. There are 3 types of cargo containment systems (CCS) for LNG vessels like LNG carriers, FSRUs and FLNG (CCNR/OCIMF, 2010), which are:

- Self-supporting moss CCS (Cargo Containment System)
- Self-supporting prismatic IHI CCS (SPB)
- Membrane type CCS

Moss tanks are a type of self-supporting CCS which are spherical in shape and are not the part of the ship's hull.

Moss tanks rest on the ring structure which is supported on the cylindrical skirt. This arrangement of moss tank is inside the tin compartment which acts as a secondary barrier outside the primary barrier of foam doused with inert gas, particularly nitrogen, to keep the moss tank dry and insulated. Natural gas detection system in the primary barrier of foam is there to detect any leakage. Moss tanks give two decisive advantages, one because of moss tank being independent of ship's hull, the tank can expand and contract under cryogenic conditions without damaging the hull, second advantage is due to the spherical design of moss tanks, they do not have sloshing constraints unlike membrane type CCS. Despite these advantages, moss CCS accounts for only 26 % of the active LNG fleet because of limited deck space for process equipment, due to their spherical design which protrudes above the deck and limits free deck space. Prismatic IHI tanks (SPB) are also self-supporting tanks and with their design offer free deck space without compromising on sloshing and the integrity of hull structure. However, currently there are only two vessels with this type of Cargo containment system, namely Arctic Sun and Polar Eagle, both built in 1993 and with a capacity of only 89,900 cubic meters. SPB tanks cannot be an option for this project. Membrane tanks unlike self-supporting form part of the ship's hull with only one primary barrier. Membrane CCS are susceptible to sloshing impacts, hence they require to be filled 70% volumetrically up to a height not less than the 10% of the tank's height. Since, FSRUs are operational throughout the year, maintaining these regulations for tank storage will become a task. Also to make membrane type CCS safe for FSRU, double row arrangement can be provided, but that increases the cost of the project significantly, so for this project LNG carrier with moss type CCS was used.

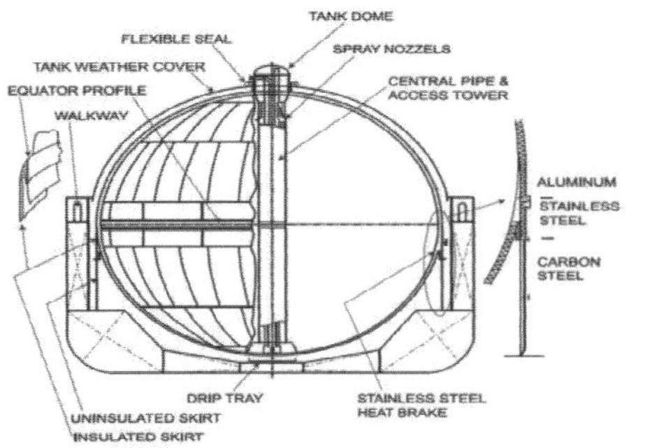

Figure 1: Spherical tank system and cargo hold configuration

4.3 LNG CARRIER SELECTION

For this project moss type LNG carrier was selected among 18 steam driven moss carriers that were laid up in 2016, with the following particulars:

Power	Steam Propulsion
Gross Tonnage	121,413
Dead Weight	75,891 tons
Carriage Capacity	152,880 m^3
Overall Length	289.53 m
LBP ×B×D	277×49×27.5
Draft	11.88 m
Block Coefficient	0.731
Number of Tanks	4

Table 1: LNG Carrier Particulars

4.4 PROCESS SYSTEM DESIGN

4.4(a) PROCESS FLOW SCHEME AND BASIC DESIGN

A typical FSRU flow scheme includes the unloading of LNG from LNG carrier through unloading arms or hoses into the FSRUs cargo tanks. The low pressure pumps send the LNG from cargo tanks to the recondenser at a pressure of about 5 bar, where LNG is contacted with the boil off gas, which gets condensed back. The high pressure pumps then take the low pressure LNG to the vaporizers at a pressure of about 100 bar g or 50 bar g, depending on the export customer (50 bar g in case of a gas power plant and 100 bar g in case of a gas network). The high pressure vaporized LNG is then metered and exported through gas export arms or hoses to the customer or pipeline. A typical Process scheme is illustrated in Figure 2.

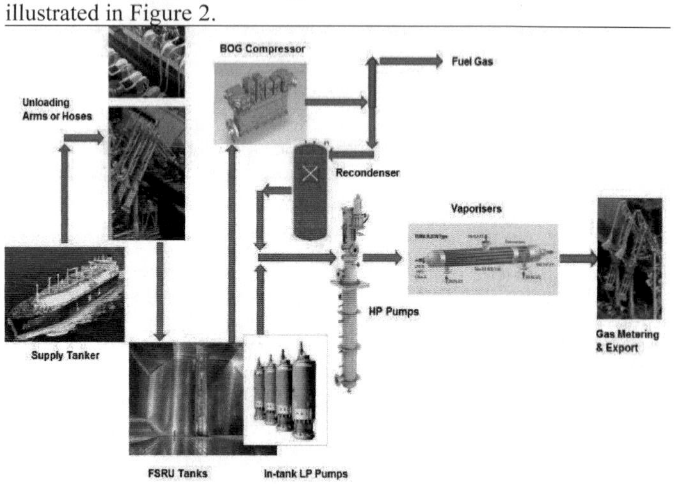

Figure 2: Typical FSRU Process System scheme

The FSRU will have regasification capacity of 2.5 MTPA/297 ton/hr and hence the send out rate will be 294.1 mmscfd (Million standard cubic feet) (1 MTPA of NG = 117.64 mmscfd). The compressors will be driven by electrical motors. Hence in this project high pressure cryogenic motor pumps are required to give a net flow rate of LNG at 294.1 mmscfd with pressure head difference of 95 bar, as the natural gas will be supplied to the gas network. Considering the Boil off gas rate of 0.16% for moss tanks (McGuire et al., 2000), the boil off gas compressors will be required to operate at the flow rate of 11.4285714 ton/day with pressure head of 5 bar. The basic LNG regasification process design is given in figure 3.

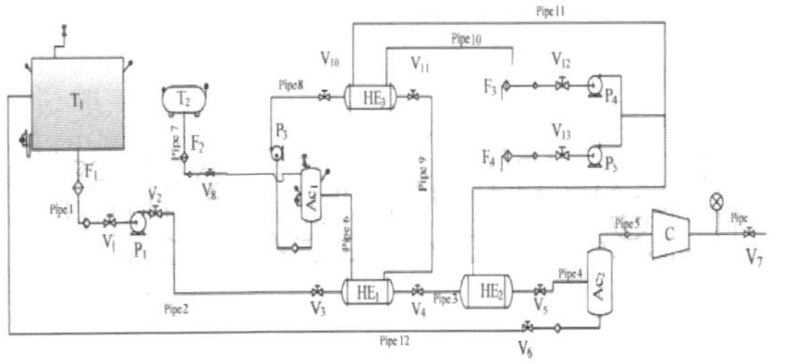

Figure 3: Regasification Process System basic design

The regasification of LNG to natural gas is done in 2 stages in cascade process in a typical regasification facility. LNG is first heated in propane compact heat exchanger (HE1) in which the temperature of LNG increases from -162 C to -10 C at about 100 or 50 bar. Since this temperature is unsuitable for the gas grid pipelines, the natural gas is further heated in shell and tube sea water heat exchanger (HE2) where the temperature is increased to 15 C. The propane used in HE 1 is in closed loop at 5 bar pressure and regasified in titanium heat exchanger HE3 by sea water. Since, the whole process involves high pressure, natural gas, Propane and critical components, a comprehensive risk assessment is advised to safe operation of the facility.
Considering that 27 C, sea water leaves the system at 20 C, mass flow rate of sea water in HE3 and HE2 in ton per hour can be calculated according to heat exchange equations and applying 0.85 as efficiency coefficient in HE3, HE2 and HE 1.
The equation is

$$q_{m\ seawater}\,(HE3) = q_{m\ LNG}\,(C_{-10} - C_{-162})\,/\,s\ (7)\ (.85)\ (.85)$$

$$q_{m\ seawater}(HE3) = 297.02375(897911.5\text{-}215875.6)\ /\ (3850)(7)(.85)(.85) = 10{,}404\ (ton/hr)$$

$$q_{m\ seawater}\,(HE2) = q_{m\ LNG}\,(C_{15} - C_{-10})\,/\,s\ (7)\ (.85)$$

$$q_{m\ seawater}\ (HE2) = 297.02375(\ 215{,}875.6 - 132{,}200.8)/\ (3850)(7)(.85) = 1085\ (ton/hr)$$

$$q_{m\ seawater}\,(Total) = 11{,}489\ ton/hr$$

$$q_{m\ seawater}\,(Through\ 1\ pump) = 5744.5\ ton/hr$$

4.4(b) CONVERSION CONCEPT, MODIFICATIONS AND LAYOUT

Marine design and process system design exercise influence on the conversion concept, the modifications and the design layout. In this project, the process systems and internal turret are placed inside the bow side moss tank after removing the moss tank and making modifications which includes the removal of skirt, hopper plate and drip tray to achieve plain bottom for the placement of equipments. Since the process equipments are placed in the bow tank, sloshing is reduced significantly, by up to 30% (Kannah et el., 2006). Weight that is removed by removing the tank corresponds to 17,000 ton and it is believed that the weight of the process equipments will be less than 17,000 ton and the net weight can be adjusted by using ballast tanks to achieve unchanged draught. The modifications are done as per the rules for main class ships (DNV, 2011). Following modifications are made to achieve a 4 deck layout for the accommodation of process equipments and turret, these include:

- Provision of double bottom with height of 2 m.
- No of girders are increased to give support to double bottom structure, such that the distance between each girder is reduced to 3 m from 5m.
- To ensure the operation of safety system even in case of accidents (DNV, 2015 b), the emergency areas like regasification are provided on the highest deck to ensure availability to personnel, in case of need and also to provide ventilation of hydrocarbons.

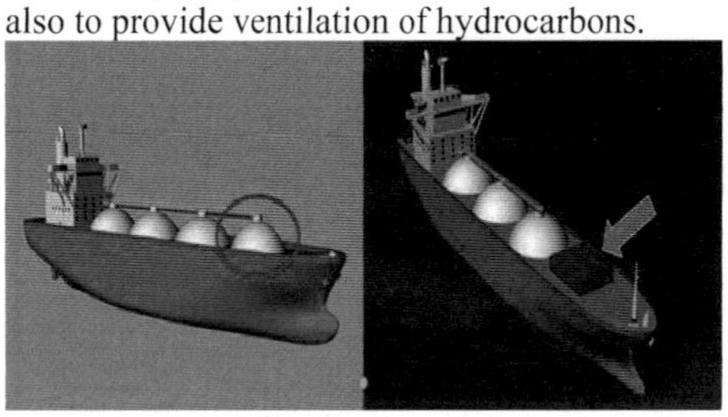

Figure 4: FSRU conversion concept

conversion structure will have a multi deck layout and will accommodate 2 sea water pumps for vaporizers, Propane compact heat exchanger (HE 1), Shell and tube heat exchanger (HE 2), Titanium heat exchanger (HE 3), 1 HP (High pressure) cryogenic motor pump, 1 propane tank, 1 compressor, Propane accumulator, Natural gas accumulator. The structure layout will consist of four decks. The vertical distance between each deck will be 3.6 meters. The first three decks will be placed inside the Containment tank, while the fourth deck will be placed above the main deck. The safest distance between the equipment is 4 meters, and between the equipment and tank boundary is 3 meters (DNVGL, 2016).

Deck	Equipment
First	P_4, P_2, P_3
Second	AC_1
Third	HE_3, T_2
Fourth	HE_1, HE_2, AC_2, C (Compressor)

Table 2: Multi Deck Equipment Layout

4.4(c) PROCESS DESIGN

Based on the process design specifications, equipments are finally selected and put onboard as per the layout and modifications established in section 4.4 (b).
The following calculations are performed to get the specifications:

- Sea water pump (1 Unit) power requirements Psea water (kW):
  - Density = 1025 kg/m3
  - Flowrate of the pump = 5604.3902 m3/h
  - Head difference (Height) = 10.8 +2 = 12.8 m

- Psea water (kW) = q ρ g h / (7*3.6 *106) (Chalmers et al., ND)

  Where
- Psea water (kW) = hydraulic power (kW)
- q = flow capacity (m3/h)
- ρ = density of fluid (kg/m3)
- g = acceleration due to gravity (9.81 m/s2)
- h = differential head (m)

Psea water = 5604.3902* 1 025*9.81*12.8 / (7*3.6*106)
Psea water = 28.624022 kW = 28624.022 W
Taking efficiency ratio of 0.85 we get
Psea water = 28.624022 kW/ 0.85 = 33.6753207 kW = 33675.3207 W
Total Power required by 2 sea water pumps for HE1 and HE3= 67.3506 kW

- High Pressure LNG pump power requirements (1 unit):

Same procedure will be followed as in the previous example, replacing the values:

- Density = 457.7312 kg/m3
- Flowrate = 297 ton/h = 648.85 m3/h
- Pressure difference = 95 bar = 9 500 000 Pa

To get the head difference estimated, we use density and difference in pressure to determine it:
Pressure (bar) = Height (m)*density (ton/m3)/10
H=10P/D = 10*95/0.4577312 = 2 075.454 m
$P_{HP\ LNG}$ = 1680 kW
Taking efficiency ratio of 0.85 we get:
$P_{HP\ LNG}$ = 1976.47 Kw

- Similarly we get the power required by 1 LP (Low pressure pump) = 78.01857586 kW = 78018.57586 W

Power required by 3 LP (Low Pressure Pumps) = 234.0557276 kW = 234055.7276 W

- BOG (Boil off gas) compressor power requirements:
- Density of NG = 0.656kg/m3
- Flowrate = 1142.85714 ton/day = 46.8 ton/h = 71.33 m3/h
- Pressure difference = 5 bar = 500 000 Pa
- Head difference = 10*5/0.656 = 76.22 m
- PNG = 0.656*9.81*76.22*71.33/ (3.6*106) = 9.72 kW
- PNG = 9.72 kW = 9720 W

Assuming efficiency of the compressor of 0.85:

- PNG = 11.43 kW = 11430 W
-

5. COST ANALYSIS

Cost of any project is composed of

- Capital Cost
- Operational Cost

The following table shows the capital cost involved in this project compared with an onshore terminal and New Build FSRU, for which values were taken from (Songhurst et al., 2017) and few were calculated.
In CAPEX, there is 30% of costs for contingency in case of onshore terminal and 10%, in case of FSRU (Songhurst et al., 2017).
The Owner's cost are the costs pertaining to decision making like support service, conceptual and impact assessment, Government NOCs, the supply contract preparation etc. which impact the FID (Final Investment Decision). On the basis of process design the equipments can be selected by researching for companies like bluewater, Ebara Pumps etc.

COMPONENT	ONSHORE (Million USD)	FSRU (NEW BUILD) (Million USD)	FSRU (CONVERSION) (Million USD)
Jetty	26	26	26
Piping (18 km × 3 million USD / km)	54	54	54
Unloading Lines	100	N/A	N/A
LNG Containment System	180	In FSRU	In FSRU
FSRU Vessel	N/A	250	120
Process Plant	100	In FSRU	65
Utilities	60	In FSRU	15
On Shore Interface Infrastructure	N/A	30	30
Total	520	360	310
Contingency	156	36	31
Owner's Cost	74	54	54
Total CAPEX	750	450	395

Table 3: CAPEX comparison between Onshore, New build and conversion

The Operating costs are the costs which are incurred during the operation of the FSRU and are cumulated every year, operational costs include personnel wages, maintenance and spare parts, insurance, services like tugs, Interest rate.
Taking operational cost per year to be 18.75 million USD/ year for onshore terminal, 11.25 million USD / year for New build FSRU and 9.875 million USD / year for converted FSRU (2.5% of CAPEX) (Rate et al., 2014)
Also 1 ton NG= 51.32 mmbtu. Taking on average Cost of LNG = 4 USD/mmbtu = 205.28 USD/ton (Sen et al., 2017) and sale price of imported LNG in India = 5.8 USD/mmbtu = 297.656 USD/ton (Sen et al., 2017)
Since FSRU in this project requires 597126.9663 mmbtu from natural gas per year as established in section 7, Cost incurred due to power consumption of the FSRU will be 2388507.865 = 2.388507865 million USD/year (4 USD/mmbtu × 597126.9663 mmbtu)
Cost of purchasing the LNG/year = 510.8113619 million USD (Gas used for power generation is deducted)
Taking tax of 2.5% on imported LNG in India (Palnitkar et al., 2017), the tax per year = 297.656 USD/ton × 2488364 ton × .025 = 18.51691187 million USD/year.

- The net OPEX for onshore terminal gets 550.4667816 million USD/ year

- The net OPEX for new Build FSRU gets 542.9667816 million USD/ year
- The net OPEX for converted FSRU gets 541.5917817 million USD/ year

Clearly cost advantage is with FSRU based on conversion which is employed in this project. Considering these, cost analysis is conducted over a 10 year period, which is the project life of this project and the sales done per year will be 297.656 USD/ton × 2488364 ton = 740.676474.8 million USD / year
We take the discount rate to be 7%

Year	Capital Cost	Operational Cost	Sales	Discount Factor	PV Operation	PV Sales
0	395	541.5917817	740.6764748	1	541.5917817	740.6764748
1		541.5917817	740.6764748	0.93	503.680357	688.8291216
2		541.5917817	740.6764748	0.86	465.7689323	636.9817683
3		541.5917817	740.6764748	0.79	427.8575075	585.1344151
4		541.5917817	740.6764748	0.74	400.7779185	548.1005914
5		541.5917817	740.6764748	0.68	368.2824116	503.6600029
6		541.5917817	740.6764748	0.63	341.2028225	466.6261791
7		541.5917817	740.6764748	0.58	314.1232334	429.5923554
8		541.5917817	740.6764748	0.54	292.4595621	399.9652964
9		541.5917817	740.6764748	0.5	270.7958909	370.3382374

Table 4: Cash Flow Analysis over a period of 10 year for the project

Hence, the net present value of 1072.588914 Million USD over 10 years, Payback period of 2 years 23 days and internal rate of return of 50.4%

6. CONCLUSIONS

In this project FSRU is acknowledged as an option to onshore regasification terminals, before the latter gets completed like before Brazil Pecem terminal, there was an FSRU and later it was shifted to another area, after the completion the terminal, which ensured the meeting of the natural gas requirements of the area in a cost effective way.
Also FSRUs are flexible way to address the natural gas requirements of the globe like after the development of shale gas infrastructure of USA, FSRU Neptune was shifted to Turkey to meet their natural gas requirements.
FSRUs also ensure fast track projects like Ain Sokna terminal was completed within just 5 months
Conversion of FSRUs can bring better economics as many small LNG tankers are on the verge of laying up, these can be brought on even lower prices and can boost cost effective natural gas import infrastructure.
Since, the modification after the conversion to FSRU are not much, these can be redeployed as LNG carrier, in case of need and back to FSRU again
This project also highlights the opportunities of gas infrastructure projects in consumer countries like China, India, Bangladesh, Pakistan, Jamaica who are committed to reduce their environmental footprint post COP 21 summit and are now encouraging natural gas by lowering the tax, in this project tax on natural in India is 2.5%, which is significantly lower compared to many countries.

8. ACKNOWLEDGEMENTS

I will like to express my appreciation and my deepest gratitude to Professor Peilin Zhou, my supervisor for his motivation and lectures in Advance Marine Engineering for giving the principles of knowledge and ideas to apply in the dissertation. I will also express my gratitude to University of Strathclyde University, UK for masters in Marine Engineering.

9. REFERENCES

1. Waters, L. B., Mueller, C. V., Hellen, P. C., & Hurst, G. L. (2007). Design and construction of gravity based structure and modularized LNG tanks for the adriatic LNG terminal. ExxonMobil Development Company.
2. Aronsson, E. (2012) 'FLNG compared to LNG carriers Requirements and recommendations for LNG production facilities and', (X). Available at: http://publications.lib.chalmers.se/records/fulltext/162630.pdf.
3. Sen, A. (2017) *India's Gas Market Post - COP21.*
4. Det Norske Veritas (2012) 'DNV-OS-C102 Structural Design of Offshore Ships', (October).
5. DNV. (2006). Fatigue Methodology of Offshore Ships. *DNV Recommended practice C206.* pp.47
6. DNV. (2010). Environmental Conditions and Environmental Loads. *Recommended practice DNV-RP-C205.* pp.47
7. DNV. (2011). Hull Structural Design, Ships with Length 100 metres and above. *Rules for Classification of Ships.* Part 3 Ch. 3
8. DNV Offshore Technical Guidance. (2011). Floating Liquefied Gas Terminals. *Floating Liquefied Gas Terminals" Report OTG-02.,* (March), pp. 1–96.
9. DNV. (2013). Conversion of ships. *Classification Notes No 8.* pp.7
10. DNV. (2015). Structural design of offshore ships. *DNV GL Offshore Standard C102.* pp-66
11. DNV. (2015b). Safety principles and arrangements. *DNV GL Offshore Standard A101.*
12. DNV. (2016). Liquified Gas Carriers. *Rules for Classification of ships.* Part 5 Ch. 5, pp.17
13. DNV Offshore Technical Guidance. (2011). Floating Liquefied Gas Terminals. *Floating Liquefied Gas Terminals" Report OTG-02.,* (March), pp. 1–96.Aronsson

14. Consumption, C. *et al.* (2017) 'BP Statistical Review of World Energy June 2017', (June).
15. British Petroleum (2017) 'BP Statistical Review of World Energy 2017', *British Petroleum*, (66), pp. 1–52. doi: http://www.bp.com/content/dam/bp/en/corporate/pdf/energy-economics/statistical-review-2017/bp-statistical-review-of-world-energy-2017-full-report.pdf.
16. Corporate (2011) 'Corporate Profile'. Available at: http://phx.corporate-ir.net/phoenix.zhtml?c=135383&p=irol-homeprofile.
17. March, F. (2017) 'Natural Gas Outlook in India', (March).
18. Franklin, D., Reeve, H. and Hubbard, B. (2010) 'Converting existing LNG carriers for floating LNG applications', *Proceedings of the Annual Offshore Technology Conference*, 2(May), pp. 1641–1651. Available at: http://www.scopus.com/inward/record.url?eid=2-s2.0-77955378775&partnerID=40&md5=ff641a64722fd98d1a94385d19b6730d.
19. Building, G. F. O. R. (2014) 'Guide for Building and Classing Floating Offshore Liquefied Gas Terminals', 2010(June).
20. Songhurst, B. (2017) *The Outlook for Floating Storage and Regasification Units (FSRUs)*.
21. Lee, D.-H. *et al.* (2014) 'Research of design challenges and new technologies for floating LNG', *International Journal of Naval Architecture and Ocean Engineering*, 6(2), pp. 307–322. doi: 10.2478/IJNAOE-2013-0181.
22. 'LNG-fuelled deep sea shipping The outlook for LNG bunker and LNG-fuelled newbuild demand up to 2025' (2012).
23. 'IGU World LNG Report' (2017).
24. 'LNG-fuelled deep sea shipping The outlook for LNG bunker and LNG-fuelled newbuild demand up to 2025' (2012).
25. 'shell-lng-outlook-2017-infographic' (no date).
26. Bluewater (2017). *Internal Turret Mooring systems – Bluewater.* [online]. [Accessed 20/03/2018]. Available at: http://www.bluewater.com/products-technology/mooring-systems/subsea-mooring-and-Unloading-point-smop/internal-turret-mooring-systems/
27. Dong-Hyun, L., Mun-Keun, H., Soo-Young, K. and Sung-Chul, S. 2014. Research of design challenges and new technologies for floating LNG. *International Journal of Naval Architecture and Ocean Engineering.* **2(6)**, pp. 307-322
28. Howell, G. B. (2006). Spread Moored or Turret Moored FPSO's for Deepwater Field Developments. In: *Proceedings of Offshore West Africa 2006*. pp. 1–21.
29. Howell, G. B. *et al.* (2006).Spread Moored or Turret Moored FPSO's for Deepwater Field Developments. *Proceedings of Offshore West Africa 2006*, pp. 1–21.
30. Jihyun, S. and Kyu-Yeul, L. (2014). Optimal liquefaction process cycle considering simplicity and efficiency for LNG FPSO at FEED stage. *Computers and Chemical Engineering*. **63**, pp.1-33
31. Kannah, T.R. and Natarajan, R. (2006). Effect of turret location on the dynamic behaviour of an internal turret moored FPSO system. *Journal of Naval Architecture and Marine Engineering*. pp.23-37
32. Lee, D. (2014). Research of design challenges and new technologies for floating LNG. *International Journal of Naval Architecture and Ocean Engineering*. **6(2)**, pp. 307–322.
33. London T, E., Arun S, D. and L Allen, Q. (2001). A Comparison Between Turret and Spread Moored F(P)SOs for Deepwater Field Developments. *Deep Offshore Technology*, pp. 1–23.
34. Wall, M. *et al.* (2002). *HSE Health & Safety Executive Failure modes, reliability and integrity of floating storage unit (FPSO, FSU) turret and swivel systems.* [online]. [Accessed 1/03/2018]. Available at: http://www.hse.gov.uk/research/otopdf/2001/oto01073a.pdf
35. Wichers, J. (2013). Guide to Single Point Moorings, *Books. Google. Com*, p. 456. [online]. [Accessed 11/02/2018].Available at: http://books.google.com/books?hl=en&lr=&id=Nz6TeBXOTTIC&oi=fnd&pg=PA13&dq=Guide+to+Single+Point+Moorings&ots=oMm-l7ofNk&sig=716xNg4CNAaxn-OZVL-n_OlN128.
36. Outlook, L. N. G. (2017) 'SHELL LNG OUTLOOK 2017', pp. 1–4.
37. March, F. (2017) 'Natural Gas Outlook in India', (March).
38. David Rowlands. 'Höegh signs agreements for FSRU conversion project'(2016). Available at: https://www.lngindustry.com/regasification/24082016/hoegh-signs-agreements-for-fsru-conversion-project-2946/
39. Chalmers, D. W., Tanamachi, M. and California, C. M. (no date) 'THE LNG INDUSTRY IS MOVING TOWARD LIFE CYCLE COST History of the LNG Pump Industry', pp. 1–15.
40. Kulitsa, M. and Wood, D. A. (2018) 'Enhanced application for FSRU recondensing

equipment during periods of low or no gas send out to minimize LNG cargo losses', *Petroleum*. Elsevier Taiwan LLC, pp. 1–10. doi: 10.1016/j.petlm.2018.01.002.

41. Sekhara Rao. (2012) 'PRE-FEASIBILITY STUDY REPORT Development of Offshore LNG FSRU Facility at Kakinada Deep Water Port , Kakinada , Andhra Pradesh', (August).
42. Jestin, N. *et al.* (2005) 'Offshore LNG receiving terminals , new architectures'.
43. Jun, L. Y. U. (2017) 'COMPARATIVE RESEARCH ON LNG RECEIVING TERMINALS', (October).
44. Valkenburg, A. Van and Dupalo, J. (2004) 'DESIGN OF A FLOATING STORAGE AND REGASIFICATION UNIT (FSRU) FOR OFFSHORE WEST AFRICA Table of Contents'.
45. Regan, T. (2017) 'Emerging Technologies for Small-Scale Grids', (October).
46. Quirijns, S. (2015) 'LNG Regasification Terminals'.
47. Smith, S. and Finley, C. (2016) 'The bigger', (February).
48. Methods, A. (1995) 'LNG CARGO T RANSFER CALCULATION METHODS AND ROUNDING - OFFS', (April 2015), pp. 1–14.
49. Waters, L. B. (no date) 'DESIGN AND CONSTRUCTION OF GRAVITY BASED STRUCTURE AND MODULARIZED LNG TANKS FOR THE ADRIATIC LNG TERMINAL, Italy .
50. Scarpa, G. and Dogliani, M. (no date) 'LNG Receiving Terminal : The floating alternative'.
51. Regan, T. (2017) 'Emerging Technologies for Small-Scale Grids', (October).
52. Martiri, V. and Milanese, S. D. (no date) 'SAIPEM SAIPEM SOLUTIONS FOR LNG SAIPEM CAPABILITIES'.
53. Wood, D. A. (2018) 'A review : Optimizing performance of Floating Storage and Regasification Units (FSRU) by applying advanced LNG tank pressure management strategies', (August 2017), pp. 1391–1418. doi: 10.1002/er.3883.
54. Egashira, S. (2013). LNG Vaporizer for LNG Re-gasification Terminal. Kobe: Takasago Equipment Plant, Energy & Nuclear Equipment Div., Machinery Business.
55. Loots, E., & Buchner, B. (2004). Wave run up as important hydrodynamic issue for gravity based. Wageningen: MARIN.
56. Cha, S.-S., Bae, G.-O., Lee, K.-K., Lee, D.-H., & Bodin, J.-L. (2007). Evaluation of drainage system around a lined pilot cavern for underground cryogenic LNG Storage . London: Elsevier.
57. Favi, A. (2014). OLT Livorno FSRU: an innovative solution for the gas industry. Livorno: Saipem.
58. P, C. E. E. L. (2014) 'Excelerate Energy : Efficient And Cost-Effective Solutions for the Global Gas Market Who We Are : Pioneer and world leader in innovative'.
59. Blackwell, B. et al. (no date) 'GOLAR LNG : DELIVERING THE WORLD ' S FIRST FSRUs GOLAR LNG : DELIVERING THE WORLD ' S FIRST FSRUs'.
60. Moreno, J. (2016) 'FLNG Concept Design', (February 2014).
61. McGuire and White. (2000)'Liquefied Gas handling principles on ships and in terminals'. SIGTTO(3)
62. Rate, C. (2014) 'USAID Energy Policy Program FSRU Tariffs Paper – Government of Pakistan', (July 2013), pp. 1–2.
63. Aronsson, E. (2012) 'FLNG compared to LNG carriers Requirements and recommendations for LNG production facilities and', (X). Available at: http://publications.lib.chalmers.se/records/fulltext/162630.pdf.
64. Palnitkar, B. U. and Aggarwal, M. (2017) 'India strengthens its commitment to infrastructure investment in 2017 budget', (February), pp. 2016–2017.
65. Travis, B. and August, R. D. (2013) 'Aeroderivative Gas Turbines', (August).

9. AUTHORS BIOGRAPHY

Rajat Bhat holds the current position of Associate Member at RINA and holds MSc degree in Marine Engineering from University of Strathclyde.

PORT DESIGN FOR SMALL LNG CARRIERS

A.P. Crowle, McDermott, UK

SUMMARY

LNG ports are traditionally designed for large LNG carriers of 125,000 to 266,000 cum capacity. This paper outlines the advances in LNG port design to combine facilities for large LNG carriers and small LNG Carriers between 1,000 cum to 80,000 cum capacity.

These loading operations for small scale LNG Carriers would take place existing large scale LNG berths and so the paper identifies any changes to the existing design that would be required in order to have the capability to load to LNG Carriers of all sizes.

The key elements of LNG ports for all capacity to date show that Small scale LNG carriers are generally equipped with an elevated manifold with 16" presentation flanges to allow standard LNG export loading arms to connect to the LNG Carrier without any risk of overreach.

Additional fenders may be required at the jetty head to allow safe berthing of LNG Carriers with capacity up to approximately 20,000 m^3.

Small scale LNG Carriers are normally not equipped with a compressor to return the Boil Off Gas (BOG) generated during loading back to shore. Therefore the differential pressure between the LNG Carrier's tanks and the onshore LNG tanks has to be sufficient to allow the BOG to free flow back to shore

NOMENCLATURE

[Symbol]	[Definition] [(unit)]
COG	Centre of Gravity
CUM	Cubic metres
ESD	Emergency shu doen
FSRU	Floating, Storage, Regas Unit
GM	Metacentre above COG
LOA	Length overall
LNG	Liquefied Natural Gas
M	Metre
MOF	Material offloading facility
MTPA	Million tonnes per annum
QRH	Quick release hook
T	Tonnes

1.0 INTRODUCTION

Liquefied Natural (LNG) terminal facilities require careful planning and analysis with regard to size of LNG Carrier, location of manifolds on the LNG carriers, technical requirements, investment costs, economic incentives, process requirements, capital investment, and other special requirements of the Government and local interest groups.

2.0 KEY ELEMENTS OF SMALL LNG CARRIERS

The following key elements apply to the port design for small LNG carriers:

- Small scale LNG carriers above 6,000 cum capacity, are generally equipped with an elevated manifold with 16" presentation flanges to allow standard LNG export loading arms to connect to the LNG Carrier without any risk of overreach. The 16" loading arms included in the large LNG carrier berth design can therefore be used to load LNG to small scale LNG Carriers and no additional loading equipment is required.
- Very small LNG carriers between 1,000 cum and 6,000 cum generally have no upper level manifolds. So loading/unloading is normally by LNG hoses.
- In combining very large LNG carriers it is expected that there will be no change to the jetty structure design required to incorporate loading of small scale LNG Carriers.

Two additional Yokohama type fenders will be required at the jetty head to allow safe berthing of LNG Carriers with capacity up to approximately 20,000 m^3. The two fenders are situated such that they are both in full contact with the parallel body of the LNG Carrier. If the Yokohama fenders are to be permanently stored at the berth, a ramp and a winch system will be required for each Yokohama fender.

- type fender for deployment and retrieval of each fender. Installation of these would not impact the design of the jetty structure but plot space would need to be allocated at the berth for storage of the fenders and for installation of the ramps and winches.
- Although the 16" loading arms can be used for loading small scale LNG Carriers, with

capacity of 6,000 to 10,000 m^3 are expected to have a design loading rate of 1,000 m^3/h which is less than 25% of the capacity of a single LNG loading arm and approximately 50% of the capacity of one LNG loading pump. When loading LNG to a LNG Carrier of this size, the flowrate should be controlled via the Hand Control Valves (HVs) on the loading arm lead lines.

- Small scale LNG Carriers are normally not equipped with a compressor to return the BOG generated during loading back to shore. Therefore the differential pressure between the LNG Carrier's tanks and the onshore LNG tanks has to be sufficient to allow the Boil Off Gas (BOG) to free flow back to shore.

3.0 REVIEW OF SMALL SCALE LNG CARRIERS

3.1 Small LNG Carriers

In order to assess the best means of berthing, mooring and loading small scale LNG Carriers, a review of the existing small scale LNG Carrier designs (between 6,000 and 60,000 m^3 capacity) was carried out.

Owners and operators of small scale LNG Carriers were contacted in order to discuss and understand their existing designs and assess the compatibility with the current berth design.

The majority of these LNG Carriers are designed for LNG bunkering purposes. As the use of LNG as marine fuel is a relatively recent development, the number of small scale LNG carriers in the 1,000 to 60,000 m^3 range is increasing as design phase shifts to the construction phase.

See table 1.

3.2 Loading Arrangement

The responses from LNG Carrier owners and operators were consistent, resulting in the conclusion that all current small scale LNG Carriers are bespoke designs with the LNG Carriers being specified to be loaded from a specific LNG export facility or facilities. The LNG Carriers are designed to be compatible with the loading arms at the specified location(s) including the height of the loading manifold and the size of the presentation flanges on the manifold.

In order to ensure compatibility with the export facility, most LNG Carriers will include an elevated manifold for loading of LNG from an onshore facility and a lower manifold which is used for ship-to-ship transfer of LNG when bunkering. Ship-to-ship transfer from the lower manifold normally takes place via hoses connected from this manifold to the receiving LNG Carrier.

The elevation of the loading manifold is such that the loading arms at the designated export facility can be connected without any risk of overreach. The manifold flanges will normally be 16" to allow connection with the standard size loading arms at the loading facility.

An excerpt from a small scale LNG Carrier (capacity of 6,000 m^3) general arrangement drawing is shown in Figure **1** 1 below to demonstrate this.

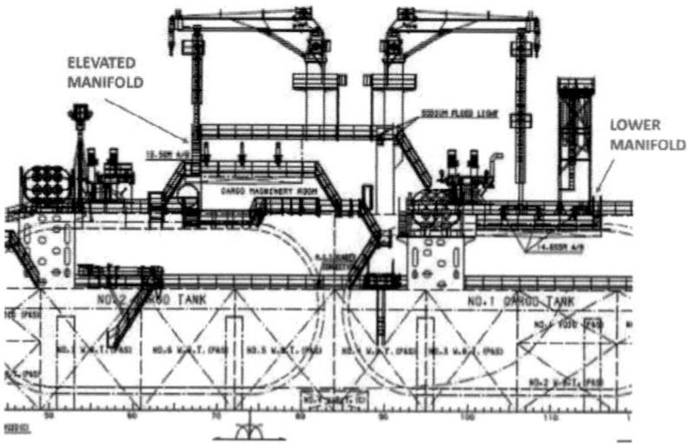

Figure 1 GA of Manifolds on Small Scale LNG Carrier

3.3 Loading rate

The design loading rate for LNG Carriers of 6,000 – 10,000 m^3 capacity is normally 1,000 m^3/h, based on responses from several operators of LNG Carriers in this capacity range. The loading rate of 1,000 m^3/h results in an overall loading time of approximately 6-10 hours (excluding time taken for connection and cool down). The design loading rate increases up to approximately 6,000 m^3/h for LNG Carriers with a capacity of 60,000 m^3, giving a total loading time of 10 hours.

The vapour return lines on the LNG Carriers are designed to handle the BOG generated at full loading rate.

3.4 Berthing/Mooring Arrangement

Due to the small size of the LNG Carriers being considered, it is often required to include additional fenders to provide sufficient contact surface on the hull of the LNG Carrier and prevent excessive pressure on the LNG Carrier's hull when the LNG Carrier is berthed. For LNG Carriers in the range 6,000 – 20,000 m^3, a maximum spacing of 60m between fenders is required to provide this contact surface.

The size of the LNG Carrier also impacts the mooring lines. Mooring of small scale LNG carriers often results in mooring lines being required to operate at different angles compared to mooring of large scale carriers. This can impact the required working angle of the mooring hooks which must be taken into account when setting the weather limitations within which loading of the small scale LNG Carriers can take place.

3.5 Other Issues

Smaller LNG Carriers are more susceptible to adverse weather conditions than larger LNG Carriers and so this must be considered when berthing the LNG Carrier.

The met-ocean conditions at the typical LNG berths are relatively benign, but the weather related berth downtime for small scale LNG Carriers will still be greater than the corresponding downtime for large LNG Carriers.

Preliminary information on typical limiting weather conditions for a standard LNG carrier (125,000 – 266,000 m^3) and a small scale LNG carrier (5,000 – 20,000 m^3) based on Northern America and Northern Europe restrictions.

During detailed design a dynamic mooring analysis and a navigation simulation exercise will be required to confirm the weather limitations for loading of small scale LNG carriers.

Small scale LNG carriers are bespoke designs and are designed to be compatible with the facility at which they are to be loaded. Therefore the basis of this paper is that for all cases, the small scale LNG carriers will be designed to be loaded with LNG via the 16" loading arms. The LNG loading rate can be as low as 1,000 m^3/h, though compared to a design loading rate of approximately 5,000 m^3/h for a single 16" arm and so the LNG loading rate has to be carefully controlled

The small scale LNG carriers loaded at the berth will have to be equipped with an elevated manifold to allow connection of the loading arms. This will require the manifold to be 13-17m above sea level.

The elevated manifold is normally provided with 16" flanges in order to be compatible with the loading arms and avoid the use of any reducers. Dependent upon the maximum loading rate of the LNG Carrier, either 1 or 2 liquid flanges will be required on the manifold, along with a single vapour flange. Since the design flowrate through a single LNG loading arm is approximately 5,000 m^3/h, a single liquid arm will be sufficient for loading LNG Carriers with capacity up to 40,000 m^3, with 2 liquid arms required for LNG Carriers with capacity of 40,000 - 60,000 m^3.

See table 2 for general guidelines on weather.

4.0 Jetty Structure

4.1 Loading Equipment

Since the existing loading arms are to be used for all cases, there is no change required to the jetty structure and no additional loading equipment is required.

4.2 Mooring Equipment

The current design of the berth has a total of 6 mooring dolphins; 3 at the fore (front) side of the LNG Carrier and 3 at the aft (rear) side. There are then 4 breasting dolphins, with 2 at the fore and 2 at the aft. Each breasting dolphin includes a cone type fender which is used to provide contact between the berth and the LNG Carrier. A photograph showing a cone fender attached to a breasting dolphin is shown in Figure 2 below.

Figure 2 Cone Fender Attached to Breasting Dolphin

- The mooring arrangement has been assessed for all 3 LNG Carrier sizes (6,000, 40,000 and 60,000 m^3) and layout drawings of the jetty have been produced with the LNG Carriers superimposed.
- The mooring points shown on the LNG Carrier are estimates at this stage and will have to be reviewed when detailed general arrangement drawings of the smaller LNG Carriers are available.
- For all mooring arrangement cases it has been assumed that the LNG Carrier will be docked pointing in a south-south easterly direction. However for smaller vessels the mooring arrangement will be evaluated as part of the terminal compatibility assessment during the chartering process. This assessment may determine the direction in which the LNG Carrier docks at the berth.

- The mooring arrangement for each LNG Carrier size will be discussed in the following sections.

4.3 Mooring of 6,000 m³ LNG Carrier

- Due to the small size of the 6,000 m³ LNG Carrier, only 2 sets of mooring lines are required; 1 set at the fore (front) of the LNG Carrier and 1 set at the aft (rear). Spring lines are then used from each of the 4 breasting dolphins to secure the position of the LNG Carrier and minimise lateral movement in the fore/aft direction.

- In order to provide sufficient surface area against the parallel body of the LNG Carrier, it is required to deploy 2 Yokohama type fenders at the berth prior to arrival of the LNG CarrierThese are pneumatic fenders with a rubber body which float on the surface of the water. They can be moved in and out of the water as required, dependent upon the LNG Carrier type that is being berthed. A photograph of Yokohama type fenders deployed at a berth is shown in Figure below.

Figure 3 Yokohama type fenders

The reason that the Yokohama type fenders are required for the 6,000 m³ LNG Carrier is that only the 2 cone fenders (attached to the breasting dolphins) on the aft side of the LNG Carrier are in contact with the parallel body of the LNG Carrier. The inner cone fender on the fore side may also be in contact with the LNG Carrier but no credit is taken for this. The Yokohama fenders provide additional surface area against the hull to maintain the LNG Carrier parallel with the berth and maintain the hull pressures within allowable limits.

The additional Yokohama fenders would be required for LNG Carrier sizes with capacity up to approximately 20,000 m³. With LNG Carriers of capacity higher than this, there would be sufficient contact between the cone fenders and the parallel body of the LNG Carrier to maintain hull pressures within allowable limits.

The Yokohama fenders would be deployed before arrival of a smaller LNG Carrier and removed from the water following the LNG Carrier's departure. Two options exist for deployment/retrieval of the fenders:

1. If there is sufficient space at the jetty head, the fenders can be stored here and lowered into the water and winched back out of the water as required. A ramp would be required for deployment and a winch for retrieval. The ramp could be installed on top of the existing jetty structure design so no change to the current jetty structure design would be required. However confirmation would be required that there is sufficient space at the jetty head to store the fenders and to incorporate the ramps and winches.

2. If there is not sufficient space at the jetty head to store the fenders, they can be stored at the harbor used to moor the tugs. The tugs tow the fenders to and from the berth as required. Prior to LNG Carrier arrival, the fenders can be secured at the berth using chains with suitable securing points provided at the berth.

See Figure 4 the berthing layout for the 6,000 m³ LNG Carrier

4.4 Mooring of 40,000 m³ LNG Carrier

Mooring of the 40,000 m³ LNG Carrier requires mooring lines from 4 of the mooring dolphins; 2 on the fore side of the LNG Carrier and 2 on the aft. Spring lines are then used from all 4 breasting dolphins.

The 40,000 m³ LNG Carrier is sufficiently large to require no additional fenders. A minimum of 3 of the cone fenders, on the breasting dolphins, are in contact with the LNG Carrier and this provides sufficient contact with the parallel body.

See Figure 5 the berthing layout for the 40,000 m³ LNG Carrier

4.5 Mooring of 60,000 m³ LNG Carrier

Mooring of the 60,000 m³ LNG Carrier requires mooring lines from all 6 of the mooring dolphins; 3 on the fore side of the LNG Carrier and 3 on the aft. Spring lines are then used from all 4 breasting dolphins.

There is no requirement for any additional fenders as all 4 cone fenders, on the breasting dolphins, are in contact with the parallel body of the LNG Carrier.

See Figure 6 the berthing layout for the 60,000 m³ LNG Carrier

4.6 Safety Systems

The Emergency shut down (ESD) systems related to LNG Carrier loading will remain as per large scale LNG loading; i.e. ESD-1 to cease loading operations and ESD-2 to disconnect the LNG Carrier from the shore.

There will be no change to the required exclusion zone around the loading area as a result of loading small scale LNG Carriers.

4.7 Fibre Optic Cable Connection

The fibre optic cable is used to provide a ship-to-shore link for operation of the ESD systems and to allow communication and data transfer between the LNG Carrier and the shore. The current specified length of fibre-optic cable connection for the project is 150m. This is expected to be sufficiently long to reach the connection point on small scale LNG Carriers and so no change to the fibre optic cable length will be required.

4.8 Jetty Gangway

The jetty gangway is used to transfer personnel between the LNG Carrier and the shore, e.g. for pre-loading meetings. For smaller LNG Carriers, the elevation of the deck will be significantly lower relative to the elevation of the jetty head in comparison to large scale LNG carriers. This may result in the angle of the gangway being too steep to allow safe transfer of personnel if the gangway is landed on the deck. Additionally the congested nature of the deck on smaller LNG Carriers can make it difficult to find a suitable location to land the gangway.

To avoid this issue, some smaller LNG Carriers include an elevated platform at the same elevation as the elevated manifold for landing of the jetty gangway. This avoids the angle of the gangway being excessively steep and also means that the gangway does not have to be landed on the congested deck area. It is recommended that small scale LNG Carriers arriving at the berth include this facility to ensure safe transfer of personnel between the LNG Carrier and the shore.

4.9 Development of Tie-ins

Since the existing loading arms are to be used for all loading operations, no tie-ins are required.

5.0 FUTURE IMPROVEMENTS

The following items are identified for potential further study work:

- Review mooring and berthing arrangement layout drawings when detailed general arrangement drawings of typical small scale LNG Carriers are available. This particularly relates to available mooring points on the LNG Carriers. This may impact the number of Yokohama type fenders required for berthing of LNG Carriers with capacity 6,000 to 20,000 m³.
- When further details of the small scale LNG Carriers to be loaded are available, a dynamic mooring analysis and navigation simulation exercise will be required to determine the weather limitations within which loading of small scale LNG Carriers can take place.
- When further details of the small scale LNG Carriers to be loaded are available, confirm the maximum LNG loading rate and the resultant vapour return rate.
- When further details of the small scale LNG Carriers to be loaded are available, confirm the pressure drop within the LNG Carrier's vapour return piping at maximum loading rate in order to calculate the required LNG Carrier pressure during loading that will allow the BOG to free flow back to shore.

6.0 CONCLUSIONS

Small LNG carriers can operate at LNG terminals design for loading or unloading LNG at berths designed for large capacity LNG Carriers, greater than 125,000cum. However if the small LNG Carrier is not in contact with 4 breasting dolphins then there are weather restrictions on operating on the LNG berth.

7.0 ACKNOWLEDGEMENTS

The author thanks his colleagues in McDermott for their assistance in preparing this paper. However the paper represents the opinion and views of the author, and does not necessarily represent those of McDermott, or any of its subsidiaries.

8.0 REFERENCES

1. UK HSE rules on consent and operational issues for LNG terminals

9.0 AUTHORS BIOGRAPHY

Alan Crowle holds the current position of principal naval architect at McDermott, London. He is responsible for naval architecture activities for offshore projects and the coordination of LNG jetty design. His previous experience includes steel jackets, semisubmersibles, FPSOs, large offshore and onshore modules, and near-shore facilities.

Small/Medium Capacity LNG Carriers in Operation

Name	Owner	LNG Capacity (m^3)	Length Overall (m)	Width (m)	Draft Loaded (m)	Year Built
Oizmendi	Cepsa	600	80	15	4	2018
Pioneer Knutsen	Knutsen	1100	73	11.8	4.5	2004
Clean Jacksonville	Toto	2200	74	15	4.6	2018
Akebono Maru	Chuo Kaiun	3556	105	17	4.7	2011
Engie Zeebrugge	GDF	5000	107.6	18.4	5.3	2017
Corialus	Anthony Veder	5800	99.7	18	5.7	2017
Cardissa	Shell	6500	119.4	19.4	5.5	2017
Kairos	Bernhard Schulte	7500	117	20	5.6	2018
Coral Methane	Anthony Veder	7500	118.4	18.6	5.7	2009
Norgas Conception	Norgas Carriers	10000	123.6	20	6.3	2011
Norgas Bahrain Vision	Norgas Carriers	12000	129	21	7	2011
Coral Energy	Anthony Veder	15000	134.4	23.5	6.8	1998
Aman Sendai	Perbadanan / NYK	18800	139.8	26	7.3	2012
Surya Satsuma	MCGC International	23096	151	26.2	7.4	2000
JS Ineos Insight	Evergas	27500	175	27	8	2015
Yuan He 1	CNOOC	30000	184.6	27.6	8.8	2015
LNG Portovenere	Exmar	65000	205	34	9.6	1996
LNG Lerici	Exmar	65000	205	34	9.6	1998
GDF Suez Global Energy	Gazocean	74100	220	35	9.5	2006
Belanak	STASCO	75000	257	35	9.5	1975
Bebatik	STASCO	75100	257	35	9.5	1972
Cheikh El Mokrani	Hyproc / MOL	75500	220	35	10	2007
Cheikh Bouamama	Hyproc / MOL	75500	220	35	10	2008
Bubuk	STASCO	77670	258	35	9.5	1975
Bilis	STASCO	77731	258	35	9.5	1975
Arctic Spirit	TEEKAY	87305	226	40	11	1993
Polar Spirit	TEEKAY	87305	226	40	11	1993

Table 1 Typical small LNG carriers

Operational Status	Criteria	
	125,000 - 266,000 m^3 vessels [1]	5,000 - 20,000 m^3 vessels [2]
Terminal open for all operations	Wind less than 25 knots	Wind speed below 20 knots, wave height below 1.0m and visibility in excess of 500m
No vessel arrivals or departures but if a vessel is at berth, cargo transfer operations already under way may continue	Wind greater than 25 knots but less than 35 knots	Average wind speed below 24 knots and wave height below 1.0m
Cargo transfer to terminate and cargo arms to be disconnected and retracted into their locked position	Wind at 35 knots	Ship leaves berth if wind in excess of 24 knots
Vessel to remain at berth and crew to deploy all other available mooring lines	Wind greater than 35 knots but predicted not to exceed 50 knots	
Vessel to put to sea	Wind predicted to increase in excess of 50 knots	

Notes

1. Criteria for 125,000 - 266,000 m^3 vessels based on USA LNG Port Guidelines
2. Criteria for 5,000 - 20,000 m^3 vessels based on Swedish Marine Technology Forum Guidelines for LNG Ship to Ship Bunkering

Table 2 Typical weather restrictions

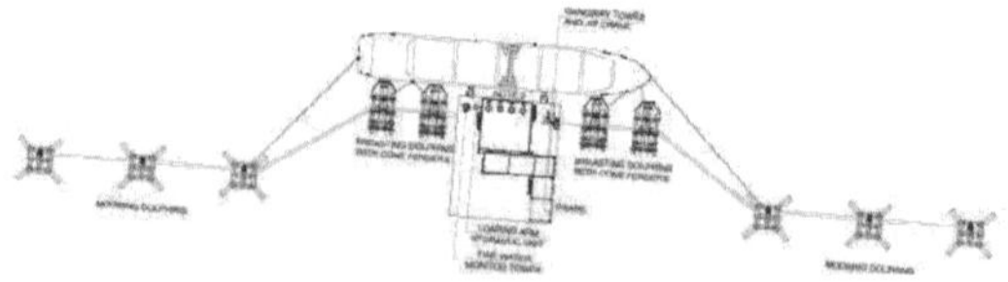

Figure 4 6,000 cum LNG Carrier, Yokohama Fenders

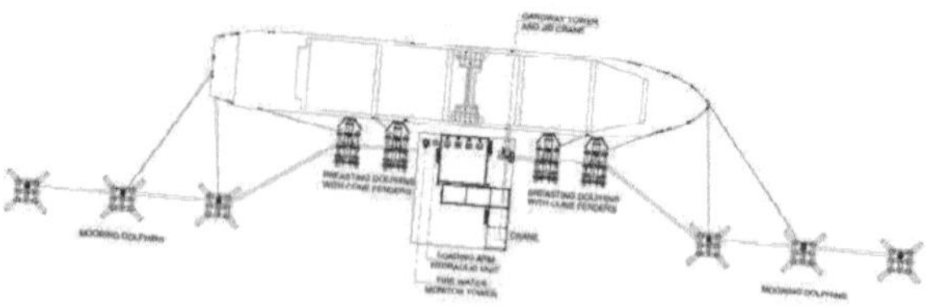

Figure 5 40,000cum LNG Carrier, on 2 breasting dolphins

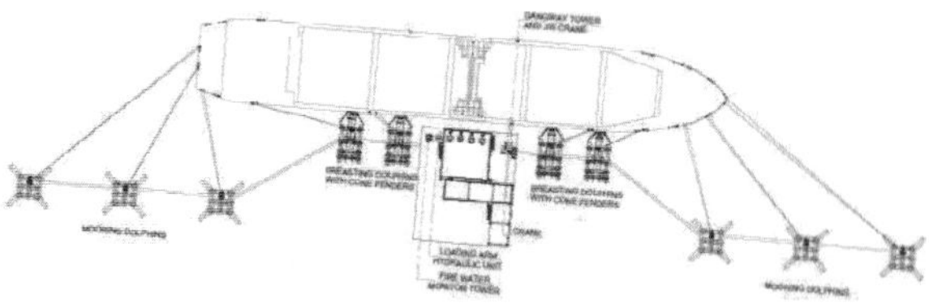

Figure 6 60,000cum LNG Carrier on 4 breasting dolphins

FUTURE FUELS FOR COMMERCIAL SHIPPING

J. E. Buckingham, BMT, UK

SUMMARY

If the size of international shipping doubles by 2050, each ship will need to emit ~25% greenhouse gases at the 2008 reference to meet the IMO requirement to halve greenhouse gases (GHG) emissions from shipping. Such a reduction cannot be achieved through the use of energy saving technologies (EST) alone, but they do have a role to play in reducing the 40% reduction in energy intensity to meet the IMO's 2030 target.

Worldwide speed limits policed by AIS-type technology can make meaningful savings and slower ships speeds greatly increase the potential benefits from wind propulsion.

However, alternative low-carbon fuel could be used to provide a large part of the solution to meet the 2030 target. This paper examines the supply of individual alternatively sourced fuels such as fuels-from-waste, methanol, ammonia and biofuels and explores the current limited production rates of their renewable sources. The paper speculates how other alternative fuel-sources may be employed in the next ten years to bridge the gap before truly zero-carbon fuels come on-line after 2030. This examination will take place in the context of shipping in the world's energy consuming sector and the way in which it may become one of the largest global consumers of liquid fuel as it declines in other sectors.

The paper seeks to create a context of how future energy for shipping bunkers will evolve within the current evolution of the re-use of existing waste streams.

1. INTRODUCTION

The IMO 2020 sulfur limit for ships bunker fuels, [1], is now in place and will lead to higher fuel prices as some ships move to distillate fuels. Such prices will encourage the take-up of Energy Saving technologies (EST) and other measures that reduce fuel consumption in the short term.

The UN Environmental programme summary report, 2019, [2] states: "GHG emissions have risen at a rate of 1.5 per cent per year in the last decade. There is no sign of GHG emissions peaking in the next few years. By 2030, emissions would need to be 25% and 55% lower than in 2018 to put the world on the least-cost pathway to limiting global warming to below 2°C and 1.5°C respectively."

The third GHG house report, [3], states the CO_2 from all shipping to be 1,135 MT.CO_2 with a total of 1,157 MT.CO_2e when other GHG such as refrigerant and air conditioning gas release are considered. This CO_2e emissions total is stated to be 2.6% of global emissions. International shipping is estimated to emit the major proportion of these sums (921 MT.CO_2 & 940 MT.CO_2e, respectively).

The estimated corresponding international shipping fuel consumption for these CO_2 emissions is 296 MT.HFO assuming all fuel is HFO (it is about 72%) at the CO_2 conversion rate of 3.114 kg.CO_2/kg.HFO.

Using sales of bunker fuels as the basis, the estimated UK contribution to shipping's GHG emissions is estimated to be 11 MT.CO_2.pa of which 9 MT.CO_2.pa is from international shipping associated with the UK. However it is to be recognised that this does not include additional fuel consumed when activity and trade-based methods are used as per Figure 2.3 of [4].

Under current UN policies, by 2050 shipping's share of a 1.5°C carbon budget will climb to 7%. The IMO MEPC 72 set out the following objectives (in summary):

a. At least a 40% reduction in carbon intensity by 2030 and pursuing efforts towards a 70% reduction by 2050, both compared to 2008 levels.
b. GHG emissions from international shipping as soon as possible and reduce them by at least 50% by 2050 compared to 2008 levels while pursuing efforts towards phasing them out consistent with the Paris Agreement.

The UN's 40% 2030 target is one element for the IMO's apparent ambitious 2030 targets to reach peak CO_2 emissions as soon as possible.

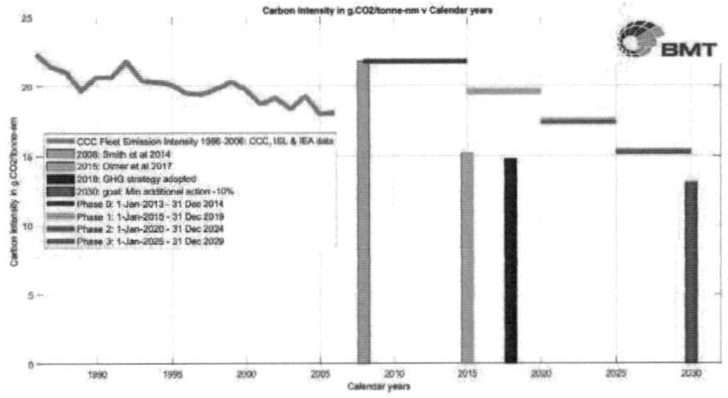

Figure 1. Carbon intensity past, present & future

Figure 1 shows how the EEDI regulations seek to reduce the carbon intensity across the board. However with a large legacy population of ships, variations in voyages, loads and speeds this is a crude measure of the progress being made. The ICCT [5] have shown how the 2030 carbon intensity target in 2015 was already 30% met due to the 2008 global financial crisis (GFC) which encouraged slow steaming due to reduced trading volumes. Based on trends in international shipping emissions and demand, combined with fleet renewal under the EEDI, shipping's carbon intensity is likely to fall by at least another 10% by 2030 without further policy interventions.

With the planned IMO-led changes to CO_2 emissions limits for shipping, there is now a greater need than ever to consider the future of fuels and the role they will play in the reaching the IMO 2050 target, [6], to halve greenhouse gas emissions (GHG) from shipping to 50% of 2008 levels as defined by the third GHG study, [7]. EST and efficiency measures alone cannot meet this target and so alternative fuels need to be considered.
However, 2050 is arguably too late for business and the industry as a whole to plan for. The world and industry needs a shorter timeframe to be able to act. The IMO 2030 target for a 40% reduction in energy intensity, [8], is therefore a good a basis upon which to make plans and this paper seeks to identify the potential solutions for alternative fuels for 2030. These studies are on-going and will be progressively refined as more information is garnered and analysed.
Whilst shipping as a business is likely to continue to grow there can be no guarantee on how future events in the next ten years will take place. With the current leadership in several large countries having indifferent attitudes to pro-active climate change measures such as GHG emissions reduction, the implementation of specific regulations at the IMO to effect the necessary drastic is not anticipated soon.

This Study

This study has considered the fuel supply and demand sides of the shipping industry over the next ten years, and addresses the range of possible changes that could occur to effect meaningful impact on the emissions from shipping. It is not a prediction but one of many scenarios that may come to pass. The formulation of this scenario seeks to identify realistic pathways to meet the IMO 2030 target.
Inevitably there has been a need to make assumptions and these are identified throughout. They will be challenged and refined as this study continues. The paper is in four parts:

- I Current situation
- II Current measures to reduce demand
- III Current decarbonisation measures
- IV Alternative fuels

2. CURRENT SITUATION

Active Ship Population

There is never a clear and accurate picture about the number of ships actively in use. This is due to the active and varied market that comprises the shipping industry as well as the difficulty in knowing what ships are laid up and in demurrage for trading purposes. The IMO third GHG report, [3], has been used as a basis for these studies with past shipping data gleaned from LR publications.
Table 1 shows the breakdown of shipping by trade for 2012 from [7] and 2019 from [9]. The latest data indicates that ship numbers have not changed greatly in the past 7 years. Whilst no ship survey is ever complete, it is assumed that ship numbers for the next ten years will also stay about the same with any increase in demand being absorbed by the current fleets and with ships being replaced on a one-for-one basis.

Ship Type	**2012**	**2019**
Gen cargo ships	16,547	16,945
Bulkers	10,094	11,562
Crude oil tankers	7,385	7,444
Chemical Tankers	4,940	5,734
Container Ships	5,130	5,152
LNG Tankers	1,609	1,980
Totals	51,730	53,629

Table 1. Aggregated number of ships by type for 2012 and 2019

Based on data from the third IMO GHG study, Table 2 shows the top ten contributors by trade and ship-size to the burden of GHG emissions from international shipping in 2012.

Ship type	**Size Range**	**Popu-lation 2012**	**Total CO_2 emissios 2012**	**Propor-tion of total**	**Cumu-lative Total**
Container	3-5 k.TEU	970	53,737	5.7	5.7
Bulk carrier	60-100 k.dwt	2,258	45,240	4.8	10.5
Bulk carrier	35-60 k.dwt	3,057	44,116	4.7	15.2
Container	5-8 k.TEU	575	42,960	4.6	19.7
Bulk carrier	100-200 k.dwt	1,245	36,340	3.9	23.6
Oil Tanker	200+ k.dwt	601	36,296	3.9	27.4
Container	1-2 k.TEU	1,304	31,015	3.3	30.7
Container	8-12 k.TEU	331	30,052	3.2	33.9
Chemical tanker	20+ k.dwt	1,475	30,027	3.2	37.1
Liquefied gas tanker	50-200cb m	463	29,283	3.1	40.2

Table 2. Ten major contributors to CO_2 emissions from international shipping

The data in Table 2 is used as the basis of considering means to reduce demand and the alternative fuel sources to be supplied. The ships in the 2012 set consumed less than the widely stated figure of over 300 MT.pa which includes the cruise sector, offshore, fishing, foot-passenger ferries, tugs and other service vessels such as dredgers.

3. MEASURES TO REDUCE DEMAND

3.1 Introduction

This sections addresses the changes that can be made on the demand side, i.e. those factors, global and industry-wide which affect the shipping industry and its fuel consumption demands.

3.2 Fuel for Vehicles

The governments of Europe & the USA are looking to reduce the number of petrol and diesel fuelled cars. In the UK, using this study's projections of current growth rates (Figure 2), there will be 5 million electric vehicle (MEV) cars by 2030, Figure 3. The growth curve looks exponential but it is being constrained by the limited production rates of EV as well as, arguably, the limited recharging points UK-wide.

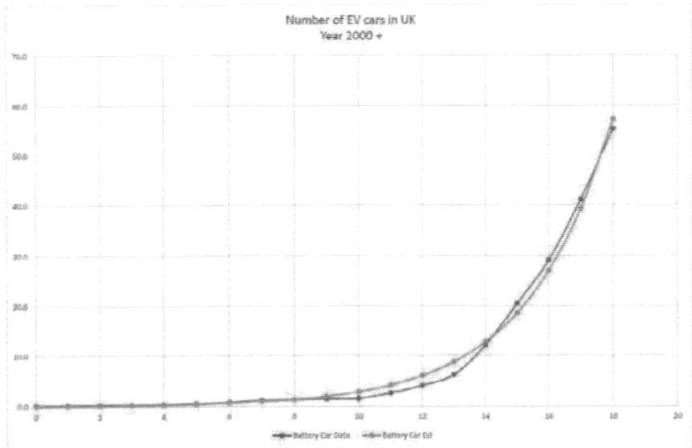

Figure2.. Number of EV in the UK 2000-2018

The UK government has a target to have 50% of cars sold in 2030 to be EV. Current annual new registrations are 2.6 MV. There are circa 30 million cars at present and this has held steady with slight growth for ten years. Assuming the total and the annual new registrations stay the same, then there will have to be 1.3 MEV sold in 2030 and using an exponential projections of recent data, this is on track. (Figure 3)

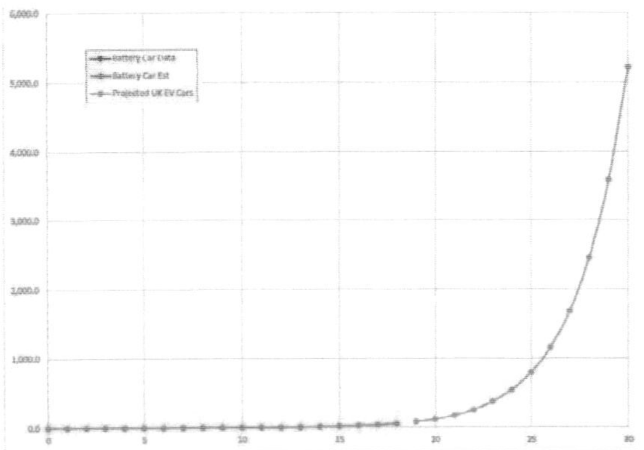

Figure 3.. Projected Number of EV in the UK 2000-2030

In 2030, a population of 5MEV amongst a total car propulsion of 30MV, represents a 17% reduction in car fuel demand.

If the 50% car fuel saving was run out throughout the EU and the USA, it would represent a significant impact on the tanker trade to those continents. In short, the complex nature of petroleum product refinement and transfer means that predicting the changes to the crude and product tanker trades will is difficult but a 10% reduction is suggested. This is also likely to hold true for light and heavy good vehicles and their growth and fuel usage. However such changes will be slow and the combined LGV & HGV UK fuel demand is about half that of the car fleet, so the introduction of electric cars will have a negligible impact on oil tanker trades up to 2030.

Domestic Heating

There are also new government policies in the EU to limit or avoid the use of domestic heating with gas and oil. This with the changes to the vehicle energy mix will affect imports of these fuels.

It could be considered that by 2030, there may be 10% fewer oil tanker and LNG carriers operations due to the changes to EV and domestic heating provisions. However as this is quite challenging to project over the rest of the EU and North America, this is not accounted for in this paper.

3.3 Battery Driven Ferries

So far, few ferries are battery powered using green electricity from shore-based supplies.

The current trend towards battery ferries is assumed to continue to 50% of ferries assumed to be battery driven by 2030 with a 30% emissions reduction for this sector.

3.4 Ship Speed Limits

The introduction of mandated ship speed limits has been mooted for many years. The feasibility of this as a means to reduce fuel consumption has been support by the President of France in 2019.

A key issue with this policy is the need for more ship-voyages to get the same material cargo from A to B in a given period. This will clearly cost more in terms of ships and crew, etc. The indicative savings were identified as the ship speeds reduce for a population of upwards from 100 vessels serve a 100nm route. Figure 4 shows how a stepwise increase in ship numbers from an initial 100 allows the vessel speed to be reduced whilst still achieving the same movement of cargo.

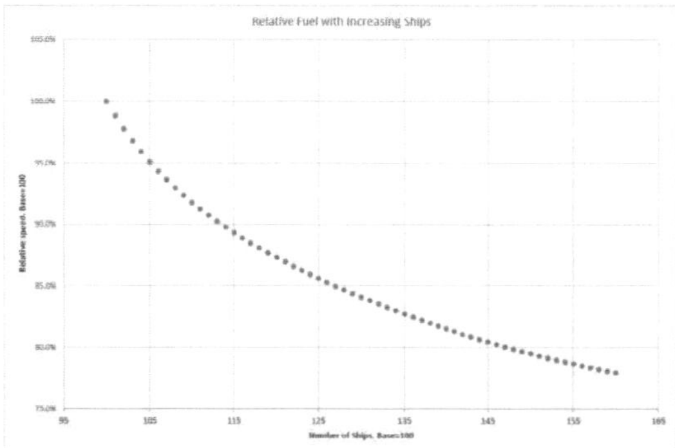

Figure 4. Ships required to service a route with decreasing speed

From this high level assessment, a 5% increase in the number of ships in service leads to a speed reduction of 5% and a 4.9% reduction in fuel consumption overall. This is assumed to be the applied rule of the sea in 2030 as a conservative measure. There are capital and operational costs associated with this.

3.5 Energy Saving Technologies

The start of a trend to introduce wind-based EST to the world's fleet will apply mainly to general cargo vessels, tankers and bulkers as they travel slowly enough to be able to gain the greatest benefit. It is assumed that LNG carriers, container ships and ferries will not be able to fit such EST. EST are assumed to give an average 8% fuel saving when using wind-EST and waste heat recovery measures such as Organic Rankine Cycles (ORC), [10].

Trade Sector	**2012 CO_2 emissions**	**Combined Factor (EST.Speed.Batts)**	**2030 CO_2 emissions**
Gen cargo ships	68,156	0.855 (0.92x0.95x1.0)	59,568
Bulkers	166,304	0.855 (0.92x0.95x1.0)	145,350
Crude oil tankers	123,614	0.855 (0.92x0.95x1.0)	108,039
Chemical Tankers	55,023	0.855 (0.92x0.95x1.0)	48,090
Container Ships	205,395	0.855 (0.95x0.90x1.0)	175,613
RoRo/Passenger ships	72,497	0.7 (1.0x1.0x0.7)	50,748
LNG Tankers	45,960	0.95(1.0x0.95x1.0)	43,662
Totals	736,949	0.86	631,069

An 8.2% fuel saving has been borne out by recent trials on the Maersk Pelican, so a combined saving of 8.0% is conservative. BMT's own experience indicates that a conservative figure is warranted due to the impact of high variability in voyage routing and weather, [11].
Table 3 summarises the impacts of these demand-side changes to the 2008 figures. This does not include reduced fuel tanker trades due to the take up of EV and reduced use of natural gas for domestic heating.

Table 3. Summary of EST Benefits

631 k.tonnes of CO_2 is 86% of 737k.tonnes.CO_2. A 14% overall saving to the international shipping fleet leads to a revised fuel demand of 191 MT.pa

4. CURRENT DECARBONISATION ACTIONS

4.1 The market

Ship usage is to be more flexible to cope with greater uncertainty in a number of areas:

- market conditions
- voyage routings
- fuel prices
- local pollution legislation

Market conditions have been unsteady and unpredictable since the GFC in 2008. Container rates have been unsteady due to a lack of consumer confidence and issues with US-Chinese trade.
Tanker rates are also unpredictable as energy continues to be an element in the world's power struggles and Asia's demand continues to grow. Additionally future energy trades may be affected by increased electrification of transport and the greater use of renewables for power generation.
In order for shipping to be more flexible to cope with these challenges, there is a need for good efficiency and flexibility across the speed range, with top-end and mid-range speeds in particular to be efficient. Propulsion equipment set-ups must be managed seamlessly with minimal operator input.

4.2 Propulsion

Alternative propulsion systems need to be configured to provide the most flexible energy-power management system. As electricity (rather than steam etc.) is the principal means of moving energy through the ships systems, a full electric or heavy-electric hybrid design is preferable to a mechanical design with a small hybrid element. However such drastic changes will not work their way through the fleet in the period to 2030 given the life-cycle replacement of most ships.

4.3 Fuels

A brief review of the set of alternative fuels most often mentioned when looking towards the 2050 objective indicates that at present they are not made in sufficient quantities, even when the carbon-associated with their supply chain is set aside.

Fuel	Chemical	Description	Energy Density	World Production	Current Capacity	Spare Capacity	
			MJ/kg	MT pa	MT pa	MT pa	MToe pa
MGO/DMA	CnHm	Complex long-chain hydrocarbons with a range of other chemicals which can leads to complex chemical emissions.	42.7	> 1000			
Batteries	Rare earth elements	Tugs, ferries, inshore vessels & short sea shipping. No good for long, high energy demands such as ocean-going ships.					
LNG/LPG	CH4 etc	Fossil fuel sourced & widely traded as a cargo. A GHG reduction is provided but not enough to meet 2050.	53.6				
Methanol	CH3OH	Made with feedstock derived from fossil fuels. Future use of renewable hydrogen.	22	75	111	36	19
Ammonia	NH3	Made with feedstock derived from fossil fuels. Future use of renewable hydrogen. Could be used in HT Fuel cells or as a dual-fuel with HFO.	18.6	150	230	80	35
Hydrogen (690 bar)	H2	Current production of hydrogen from renewable sources is limited but is growing . Prices comparable to current fuels. Current H2 used to make NH3.	142	57	0	57	190
Biodiesel	Mixed feedstocks	Mostly made from vegetable oil (22.7 BL) & palm oil (13.4 BL). Land-use 24.2+0.85 Mha=25Mha.		29		29	-
Biodiesel	Wood-based sources	Translation : 2 x wood to fuel. Licella,	36	10		10	8
Synthetic Fuel		Wide variety of syn fuels at present. Use of solar & wind to make H2 still in small quantities. H2 to be used as a feedstock to create fuels.	42.7	12		12	12
Sums							263

Table 4. Summary of Current Alternative Fuels (2019)

Table 4 shows that the total spare manufacturing capacity of current alternative fuel candidates is not enough to meet the need of shipping. The current ammonia commodity is derived by steam-based methane reforming (SMR) which is itself highly energy intensive and which consumes about 70% of the current hydrogen production of 70MT.pa, which is itself not from green renewable sources.

4.4 Hydrogen

Only 2% of hydrogen is made by electrolysis derived from renewable energy sources (IEA report , [12]). The global production of 70MT.H_2 consumes 330 MToe of energy and emits 830MT.CO_2 emissions pa, i.e. its LCA is over 11kg.CO_2/kg.H_2.

The fossil fuel energy consumption is directly comparable to the shipping industry and so any migration of ships to hydrogen bunkers without the transfer to green hydrogen will not improve the overall balance of CO_2 emissions, as 70MT.H_2 has less energy potential (233 MToe) than the 330 MToe fossil fuels used to make it.

The Norwegian electrolyzer company NEL believes there is scope for the EU to implement a H_2 production capacity of 2.8GW by 2025. Allowing that to reach 5.0GW capacity by 2030, this would allow 0.8 MT.H_2 to be produced, equivalent to 2.65 MToe. Currently 60% of the rise in wind and solar PV is from Asia,[13] but looking further afield, the IEA suggests by 2040 Australia could produce the equivalent of 100MT of oil in hydrogen per year [14]. This would be turned into ammonia and shipped to Japan and South Korea at a price of \$3/kg. This is approximately equal to a FO price of 900 USD/tonne.

Hydrogen can be bought at energy prices comparable to that for the national grid (3\$/kg) but it has a high safety risk and there is therefore an overhead capital cost (capex) for safety equipment. There is little or no land-based infrastructure for the transport and supply to ships. Onboard facilities require very low temperature cryogenic equipment robustly designed to meet the marine shock environment. There is no common technology for hydrogen as a bunker fuel and engines of all sizes and rating have little experience with the fuel. Its small molecules can easily escape seals and leads to the risk of explosive pockets of the gaseous fuel.

Clearly, there is little prospect for green ammonia and hydrogen to be used as bunker fuels in international shipping in the next ten years, apart from on prototype vessels.

4.5 LNG

The current use of LNG fuel in short sea and long-route shipping is small, but it is established and growing partly due to its low sulphur content and competitive price. Fewer than 300 ships have been or are being built with LNG as the primary bunker fuel, but more and bigger ships are being planned.

Current LNG is comprised of a range fossil fuel gases from methane to propane, but is primarily composed of methane which is a GHG with a Global Warming Potential (GWP) of over 28 on a 100 year basis. However on a 20-year basis, the timeframe of interest for 1.5°C, this rises to 84 times. Clearly there are issues with any leakage of methane as per methane slip which is recognised as an issue for some engine types.

When burnt, LNG has a ~20% lower CO_2 mass compared to longer chain fuels, but taken alone, is not a feasible solution to the GHG challenge. If Carbon Capture & Storage (CCS) can be employed, then there is promise and the re-use of the cold fuel storage may allow liquefaction of the CO_2 in the exhaust.

4.6 Alternative Fuels

The 2050 IMO target of a reduction of 50% GHG relative to 2008 emissions requires a transformation in the bunker fuels. A new fuel solution is required which has a Life Cycle Assessment (LCA) result which is better than the current fossil fuel based solution. Solutions which are currently receiving much attention such as ammonia and methanol currently require more energy to make and transport and therefore emit more GHG overall than the current fossil fuels.

4.7 Methanol

Methanol is widely used as a fuel for heating and has low sulphur and PM emissions. NOx emissions are down by 60%. It has safety hazards but these are manageable. However its calorific density is about half that of current marine fuels. The Stena Germanica is an example of a working methanol ferry and its performance is reported to be good.

Worldwide distribution and availability of methanol is also good and it is widely traded (one of five popular seaborne commodities) and has been carried as a chemical feedstock product for over 50 years. The Chinese market comprises 50% of global production and consumption with relatively little or no production in Europe. The use of methanol as a bunker fuel is being supported by the Swedish government, principally for short sea shipping as it reduces emissions of PM by 95%, SOx by 90% and NOx by 60%. It reduces CO_2 emissions by 25% at point of use.

Such fuels have low sulphur emissions but they needs significant health & safety measures to avoid vapour exposure to crew and passengers and special fire protection measures. 90% of methanol is currently manufactured by non-renewable processes which produce almost three times the CO_2 per unit energy than marine gas oil.

In 2013, only 1% of methanol was made from renewable sources. Besides the production challenges there are infrastructure challenges too; methanol's volumetric energy density is only 40% that of diesel, so clearly there would have to be more regular bunkering and therefore widespread availability.

At present, such fuels are not a feasible solution of the GHG challenge unless CCS is employed or they are manufactured using renewable energy, so called electro-fuels. They will be costly and not cost effective for shipping in the short-to-medium term.

4.8Ammonia

Ammonia can be burned in modified two-stroke engines and trials are being undertaken in Japan to assess the overall, impact on emissions and performance, [15].
Ammonia is also made through energy intensive Haber-Bosch process, usually using natural gas as the feedstock. Using the SRM and the Haber-Bosch process to combine the N_2 and the H_2, the ammonia has an LCA of about 0.85kg.CO_2/kg.NH_3. This can be halved using renewable methods, [16]. Almost half of current ammonia production is used for urea and fertiliser manufacture.
As with methanol, ammonia can be made with hydrogen derived from renewable energy and has the benefit that it does not carry a carbon atom and so cannot from a carbon oxide molecule. Ammonia also has significant H&S challenges but as with methanol it is widely traded as a sea-going commodity and so there are already storage and handling measures in place from the IMO and classification societies.

4.9 Biofuels

Biofuels (BF) continue to be trialled with commercial ships and indeed navies have often used biofuel blends when they embark Marine gas oil, which is often contaminated with automotive BF. BF has little or no sulphur content but has Fatty Acid Methyl Ethers (FAME). This makes it very vulnerable to microbiological contamination and miscible with water. BF are still made in small relatively amounts and biodiesel is an even smaller component. Ten years ago Jatropha was a promising feedstock source which could be grown on marginal land. However yields were lower than touted and the measures and time to grow it as a crop were too challenging. When the issues with land-use and sustainability are included BF is considered to be unlikely as a future bunker fuel

4.10 Electro-Fuels

Synthetic fuels can be created from a wide variety of feedstocks which are themselves created using ‘green’ hydrogen derived from renewable energy electrolysis which is then combined with other chemicals to create the feedstocks. Such fuels can be called electro-fuels, [17]. They promise to provide high quality fuels (i.e. low oxygen content etc). These will be highly sought after by the aviation industry who also need to decarbonise but cannot afford to compromise on fuel quality. Such fuels will be very costly and will struggle to be competitive apart from special areas. One case may be the creation of green-ammonia from remote high-solar, low-population places such as Australia. This fuel can then be shipped to Asia and also used for ship bunkers.
Future LNG may be from BF and future ammonia and methanol bunkers may be made using renewable energy, but the facilities to make this in any significant volume are far from being built and operational.
From the author’s personal perspective, the best way forward to preserve the current liquid-based fuel supply system for ships bunkers is to gradually blend in fuel from other sources.

5. OPTIONS FOR FUTURE FUELS

5.1 INTRODUCTION

The baseline alternative fuel options do not appear to offer a credible source of low-carbon bunker fuel for shipping in the period up to 2030. These studies have sought to identify options for change whereby drop-in fuels for assorted sources can be used to contribute to a lower carbon footprint for shipping.
The principal alternative options considered here are:

a. Fuel from forest growth
b. Fuel from forestry wood waste
c. Fuel form processed waste wood
d. Fuel from wood fuel
e. Fuel from wood pellets
f. Fuel from waste plastic
g. Fuel from household waste.

5.2 Fuel from Forest GROWTH

The consumption of 350MT.pa of fuel consumed by ships bunkers emits 1,089 MT.CO_2.pa. At average CO_2 annual absorption rates for forests of 74 tonnes.pHa.pa, this requires an additional 14.7MHa, or half as many trees again as there are in Sweden. There are no co-ordinated plans to increase forestation at this level and with so many fires in forested areas, this is a challenge for a 2030 target.

5.3 Fuel from forestry Wood Waste

There is a broad industrial drive to move the now established means for converting wood waste from forestry processes, pulp and paper factories as well as waste plastic to create wood oil. The following companies are a few of those leading this change:

- SEKAB of Sweden
- Licella of Australia

The EU funded project Rewofuel is allowing groups of companies to work together to create end-to-end processes that ensure their whole supply chain becomes economically viable.
The hydrogen to carbon ratio of lignin-cellulose is 0.2 and that of lignin is 0.4. These are far from the required value of 2.0 or more for good fuel for aviation applications. The thermal, high pressure treatment offer by modern methods leads to shortened molecular chains which are better as diesel-type fuels. The aviation industry is already heavily involved in supporting research that will offer it drop-in fuels. The recent prize given by British Airways to the following universities underlines the interest within the UK:

- Heriot-Watt
- UCL (from Waste)
- London School of Economics

Worldwide there are 1,874 MC (million.m^3). pa of industrial roundwood produced each year. Assuming a waste rate of 5%, 94MC.pa (68 MT.pa) of unusable wood can be used to create 34 MT.pa of wood oil.

On a UK basis it is estimated that 2MT.pa of excess wood waste can be utilised for domestic heating or for fuel generation.
Yet to be fully quantified is the potential use of sustainable crops such as willow and miscanthus grass (both about 19 MJ/kg) for liquid fuels.

5.4 FUEL FROM PROCESSED WASTE WOOD

The UK creates 4.3 MT.pa (2012) of wood waste arisings, i.e. municipal wood which ranges from good quality off-cuts through to chipboard and medium density fibreboard through to ex-furniture, fencing, railway sleepers, etc. Such wood does not represent a steady supply for processing but it is dry, cheap albeit with contaminants such as paint and preservatives. It usually has a calorific value of 16MJ/kg if moisture content is 15%.
On an EU scale, there could be as much as 33MT.pa which if 50% could be processed in 2030 could yield 5 MToe.pa.

5.5 Fuel from wood fuel

Worldwide there are 1,348 MT.pa of wood fuel consumed both for heating and for cooking. The use of such fuel is often from distributed non-centralised sources and the burning of this fuel can lead to black carbon and pollution issues for those who use it.

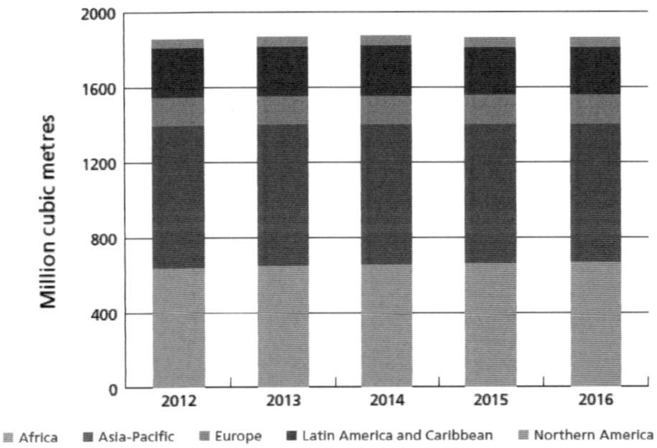

Figure 5. Worldwide Wood Fuel Production in MC.

Assuming measures could be brought in to provide electric heating and cooking in those nations where wood fuel predominates such that there was a 20% re-allocation of fuel to processing.

Africa

On a more local level, much of Africa's forests are not certified as renewable. Therefore, efforts here are best spent reducing wood-fuel usage through schemes such as those below which offer less wood-fuel intensive cooking methods or avoid wood through using gas instead of charcoal which preserves forests. Solar power is also an obvious way to replace wood for lighting.
Changu Changu Moto Fuel-efficient Cookstove
Smart gas cooking seeks to break African cities' dirty charcoal habit
Africa Goes off the Grid to Bring Power to Rural Villages
In Africa, 90% of the roundwood harvest (48 MT.pa) is used to make wood-fuel or charcoal, mostly for people to cook with. If this could be reduced by 50% 24 MT.pa of forest could be saved, equivalent to 12 MT.pa of fuel.

Europe & North America

Where wood is from certified renewable sources, there is more scope for reducing wood usage for fuel, charcoal and pellets. North America harvests 56 MC.pa of roundwood for wood fuel and the EU 149MC.pa. If 50% of this crop was used for making wood-based fuel, then it is estimated that ~10 MT.pa of black liquor crude could be created for use as a drop-in fuel.

5.6 Fuel from Wood Pellets

The production of wood pellets has grown at a rate of 2 to 3% pa. This has been driven by policies such as the EU renewable energy laws which encourage the use of renewable wood for power and heat production. In some cases, the pellets may incur many miles of travel and transport energy consumption by sea and road to where they are finally burnt in a power station. Using pellets to make liquid wood fuel local to the forest, avoids the pellet transport life cycle burden and allows usage as bunker fuels at local ports, thus reducing tanker imports to counties such as Sweden and Canada.
The current global wood pellet production is still relatively small at 29 MT.pa. If processed to form liquid fuels, they could yield 3 MT.pa of bunker fuel and save transport energy elsewhere.

5.7 Fuel from waste plastic

In addition to the UCL study above, the Swindon-based company, Recycling Technologies, shows how plastic waste can be transferred into a useful component such iso-butanol which is useful as a baseline chemical for many processes. The company is committed to addressing the circular economy and does not seek to offer its recovered product as fuel for energy processes, preferring instead to use the materials to replace new-made plastics.
However, to act quickly to address shipping emissions and to manage the waste plastic, using plastic-derived fuels such as butanol ($C_4H_9.OH$) as a drop-in fuel will reduce the dependency on fossil fuels. Butanol has a good H/C ratio but its oxygen content limits its net heat of combustion (and thus the temperature which is good for low NOx emissions. It has a flash point (FP) temperature of 28.89°C which means it has to be mixed with standard fuels to achieve a 61°C FP.
An Australian company, Licella, has created a UK company called Mura with Armstrong Energy. The UK-based business Wilton treats 20,000 tonnes of end-of-life plastic using the Cat-HTR (Catalytic Hydrothermal Reactor) process to create useful chemicals.
About 800 litres of fuel (butanol-like) can be gained from each tonne of plastic waste. The world creates 359 MT.pa of plastic per year, if 50% of this is wasted then 115 MToe.pa can be generated from it. The EU creates

15 MT.pa of waste plastic, 50% of which would yield 10MToe.pa.

5.8 Fuel from Household Waste

Of that waste which is not collected as plastic, there are plans in the UK to treat up to 600,000 tonnes pa of waste at the planned Altalto facility in Immingham, UK, [18]. This would yield 100,000 tonnes of fuel pa some of which is destined for aviation as the project is part sponsored by British Airways. As the UK generates 26.9 MT.pa of household waste this would yield 4.5MToe.pa and if applied EU wide this would yield 35 MToe.pa. Allowing for a slow uptake by 2030 as such facilities are costly to build and take up large areas, assume 10MT.pa of fuel could be provided.

5.7 Summary of Alternative Fuels

This section has identified credible sources of alternative fuels potentially for the shipping industry. There will be competition for such fuel from aviation and other industries and of course, spot fuel prices around the world and the introduction of regulation and carbon taxes will all affect how this fuel finds its way to ships' bunkers.

Table 5 give a summary of the potential fuel sources.

Fuel Type	Made oil MT.pa.2030	LCV Std fuel =42.7MJ/kg	MToe.pa
Fuel from forestry wood waste	34	33 MJ/kg	26
Fuel from processed wood waste (EU)	5	33 MJ/kg	4
Fuel from wood fuel (EU & North America)	10	33 MJ/kg	8
Fuel from wood pellets	3	33 MJ/kg	2
Fuel from plastic waste.(EU)	10	36.1 MJ/kg	8
Fuel from EU household waste	10	33 MJ/kg	8
Total			56

Table 5. Summary of potential annual alternative fuel generated in 2030

The benefit of shipping bunker fuels is that they are able to accept fuels that have a greater oxygenation than aviation fuel would accept. They therefore have a shorter pathway to direct usage and consequently hold a cost advantage. One example is butanol which has a hydrogen to carbon ratio of 2 which is comparable to diesel fuel and is considered too oxygenated to be used in aviation fuels.

6. CONCLUSIONS

The widespread use of EST, speed reduction and batteries for short-range ferries would reduce the international shipping fleet fuel demand from 223 MT.pa to 191 MT.pa by 2030. Measures using liquid fuel derived from EU and North American household waste, plastic waste, wood waste and pellets has the potential to produce an annual amount of bunker fuel of 56 MToe.pa, or 27% of the revised fuel demand. Clearly the balance of changes to shipping operations and ship designs and the use of waste and wood from renewable forests is challenging both politically and in terms of rate of change to infrastructure. Changes will need a regulatory and co-ordination role that only inter-governmental agencies can offer.

The ideas presented in this paper will continue to be reviewed as part of the VTAS project supported by BMT and Black & Veatch.

8. ACKNOWLEDGEMENTS

This paper is an output from the low carbon shipping studies being carried out by BMT as part of the VTAS joint venture with Black & Veatch. The funding and support of BMT to the author is gratefully acknowledged.

It is BMT's intention to claim copyright for this work. The kind permission and resources granted to the author by BMT are acknowledged with thanks. All findings, ideas, opinions and errors herein are those of the authors and are not necessarily those of BMT Defence and Security UK Ltd.

9. REFERENCES

1. IMO 2020 MARPOL Annex VI, Sulfur reduction agreement. IMO's MEPC 70th session in 24 – 28 October 2016.
2. UN Environmental programme Emissions Gap Report 2019. Executive summary report
3. IMO Third IMO GHG Study report . 2014
4. CCC Scope of Carbon budgets. Statutory advice on inclusion of aviation and shipping. April 2012.
5. ICCT. Policy Update. "The International Maritime Organization's Initial Greenhouse Gas Strategy". April 2018
6. IMO 2050 agreement. Adopted 13 April 2018.
7. IMO Third Greenhouse Gas Report, 2012
8. IMO 2030 target for a 40% reduction
9. Statista data for 2019 ship numbers
10. Buckingham & McCracken "Supercritical organic Rankine cycle yields useful power & emission benefits". INEC 2016.
11. Buckingham, Pearson & Storey. "Ship-data driven propulsion models & wind-based propulsion technology". RINA Wind propulsion October 2019.
12. IEA Report. "The Future of Hydrogen". Seizing today's opportunities. June 2019.
13. IEA World Energy Outlook Report 2018

14. International Energy Agency's World Energy Outlook 2018.
15. Motorship, "J-ENG to investigate hydrogen and ammonia as combustion fuels". 24-Sep 2019
16. Yusuf Bicer, et al. "Comparative life cycle assessment of various ammonia production methods" Journal of Cleaner Production, Volume 135, 2016, Pages 1379-1395, ISSN 0959-6526.
17. Mariliis Lehtveer, Selma Brynolf, and Maria Grahn. "What Future for Electrofuels in Transport? Analysis of Cost Competitiveness in Global Climate Mitigation". Environmental Science & Technology 2019 53(3),1690-1697 . DOI: 10.1021/acs.est.8b05243
18. Altalto facility in Immingham, UK.

9. AUTHORS BIOGRAPHY

John Buckingham CEng FIMechE, is the Chief Mechanical Engineer of BMT Defence & Security UK Limited, Bath, UK. John was the Chief Technologist for the ETI-funded Vessel Technology Assessment System project. John leads studies into alternative fuels marine power and propulsion, air-conditioning and hydraulics within the company and has published several papers on these subjects. Since 1999, John has been involved in a range of studies to assess the utility of EST for the commercial and military marine.

EVALUATION OF LNG CRYOGENIC ENERGY UTILIZATION IN SEAWATER DESALINATION ON LNG FUELLED SHIPS

M. Liu, School of Engineering, Newcastle University, UK
D. Wu, Department of Mechanical Engineering, The University of Birmingham, UK

SUMMARY

In the marine environment, the massive amount of LNG cryogenic energy is wasted during the re-gasifying process. Seawater desalination is one of the preferable applications for the LNG cryogenic energy recovery on board. Among all the desalination technologies, the atmospheric water generator (AWG) and single-stage conventional distillation (SCD) are selected to be applied on LNG fuelled ships. Therefore, in this study, LNG is considered as the cooling medium to condense the water vapour in the moist air due to its low temperature. The feasibility of the LNG integrated with AWG and SCD are investigated in Siemens LMS Imagine. Lab Amesim separately. The performance of the two models is compared with the conventional cooling medium systems. The efficiency and the LNG cryogenic energy consumption of the two proposed systems have been promoted at least 40% in the condensation process.

NOMENCLATURE

A	Area, m^2
ah	Absolute humidity, kg/kg
H	Specific enthalpy, J/kg
h	Heat transfer coefficient, W/ ($m^2 \cdot$ K)
hv	Water latent vaporization, J/kg
l	Characteristic length, m
$\dot{m}$	Mass flow rate, kg/s
Nu	Nusselt Number
q	Heat flow rate, kJ/kg
T	Temperature, K
Abbreviations	
AWG	Atmospheric Water Generator
LNG	Liquefied Natural Gas
RR	Recovery Ratio
SCD	Single-stage Conventional Distillation
Greek Letters	
λ	Thermal conductivity, W/ (m $\cdot$ K)
δ	Thickness of the wall, m
Subscripts	
c	Condenser
co	Cooling medium
cond	Condensate
da	Dry air
f	Freshwater
in	Inlet
lw	Liquid water
ma	Moist air
max	Maximum
out	Outlet
sat	Saturation condition
w	Wall

1. INTRODUCTION

As fewer emission of carbon dioxide and average operating heat rate for 1-kilowatt-hour electricity [1], natural gas is considered as the best substitute for coal and petroleum. The abundant natural gas resources widely distributed around the world reveal a big potential in natural gas utilization technology. When the natural gas is cooled at -162 °C, liquefied natural gas (LNG) has become the best phase condition to transport and store natural gas in smaller occupied volume. In order to promote the efficiency of LNG utilization, LNG has been considered as fuel for transportations since 1997. As recorded, until 2025, between 13 and 1963 LNG fuelled ships and 653 basing on the base case scenario will be built for deep-sea shipping as Lloyd's Register (2012) predicted [2, 3]. When LNG is converted to NG in an LNG vaporizer for the marine engine, the required heat for LNG phase changing which is called cryogenic energy [4-6], around 830 kJ/kg, should be recovered for other applications such as separation process, seawater desalination, cold storage systems and power generation rather than wasting [5].

Freshwater is one of the essential elements for industrial and human purposes especially in marine environments. In general, two kinds of desalination methods are commonly used: thermal distillation and membrane technologies. The thermal distillation can be divided into four kinds: single-stage conventional distillation (SCD), multi-stage flash distillation, multi-effect distillation and vapour compression. And the membrane technologies can be divided into three kinds: reverse osmosis, electrodialysis and membrane distillation. Apart from these technologies, freeze desalination, atmospheric water generator (AWG) and humidification-dehumidification are also proved to be workable technologies to obtain freshwater in small scales. Based on the principles of these technologies, the SCD and AWG are the preferable methods for desalination on board due to the infinite seawater and moist air resource and simple designs.

SCD explained a conventional distillation process that the seawater is flashed and evaporated under the vacuum conditions, and then the water vapour is condensed in the condenser into freshwater. As a widely-used desalination method on board, to improve the efficiency of the system is essential. One of the methods is to improve the condensation efficiency. Kumar, et al. [7] realized that the temperature difference between the surface and deep seawater can be used for flash evaporation desalination. According to the experimental result, with the coolant temperature increasing from 8°C to 15°C, the freshwater yield decreased from 5 L/h to 1 L/h due to the reduced heat removal rate from the condenser. Therefore, to further improve productivity, a lower temperature cooling medium such as LNG is a preferable option.

Generally, AWG technologies could be processed into two methods: water vapour concentration by desiccants (absorption/adsorption) and surface cooling. The first method is to concentrate the water vapour using desiccants. Although this method has benefits of energy-efficient and environmentally friendly, the higher regeneration energy, higher operating costs, maintenance and longer time to respond characteristics limited the water production and system efficiency. The second method is that the water vapour in the moist air can be extracted by cooling to the temperature lower than the dew point of moist air. As explained in Figure 1, this cooling process is explained with temperature distributions in the psychrometric chart. When the surface/wall temperature is lower than the dew point of the given moist air, the condensation occurs. Therefore, a large amount of cooling energy is required to meet higher temperature and relative humidity.

Due to the characteristics of environmentally friendly, safe storage, plenty and economical resource, natural gas has become the best substitute for coal and oil. After cooled at around -162°C, LNG occupies volume 600 times smaller than natural gas for shipping and storage which offers a better method to transport LNG to the markets all around the world. Therefore, LNG fuelled transportation has been trending in the last decade of years. In this process, the LNG vaporizers are operated to convert LNG to natural gas for engine uses. Amount of release heat, around 830 kJ/kg, during this process called cryogenic energy is preferred to be recovered for other applications such as separation process, seawater desalination, cold storage systems, waste incinerator, cryogenic CO_2 capture and power generation rather than wasting. In the desalination field, LNG cryogenic energy is always considered as a free secondary refrigerant to support freeze desalination. However, this large amount of cold energy has never been used as a cooling medium to promote the condensation process. And the freeze desalination method is not considered as a freshwater generator method on board due to its limited desalinate ability and the small scale.

In order to reduce the energy consumption and the costs, renewable energy with infinite amount: liquefied natural gas (LNG) is considered as the alternative cooling medium. In this study, the LNG cryogenic energy integrated with these two desalination methods: SCD and AWG are proposed on board. The two models are first developed in Siemens LMS Imagine. Lab Amesim and validated by experimental measurements. Then the freshwater yield of these two models is compared with the conventional cooling water system performance under the same operating parameters.

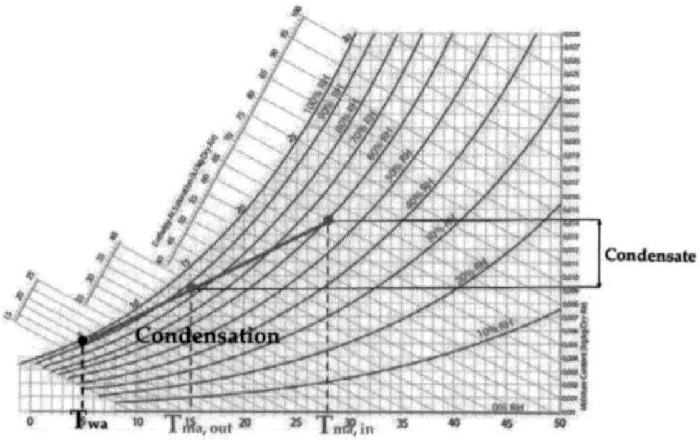

Figure 1. Psychrometric chart of the condensation process.

2. SYSTEM DESCRIPTION

The models are developed to evaluate the transient performance of these two desalination systems assisted by LNG cryogenic energy. The design of the proposed AWG-LNG system is illustrated in Figure 2. The moist air is pumped into the exchanger by the fan and then exchange the heat with the input LNG. The sub-zero temperature LNG absorbs the heat from the high-temperature moist air re-gasified into NG for the combustion engine. In this process, the water vapour is extracted from the moist air and condensed into the freshwater by the waste LNG cryogenic energy. And the design of SCD-LNG system is in Figure 3. The feed seawater is flashed into water vapour in the flash evaporation chamber under vacuum conditions. Then the water vapour is condensed into the freshwater by LNG cryogenic energy.

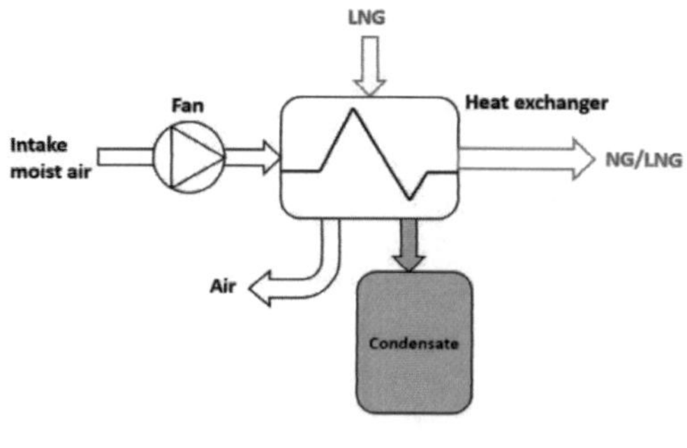

Figure 2. Block diagram of AWG-LNG system.

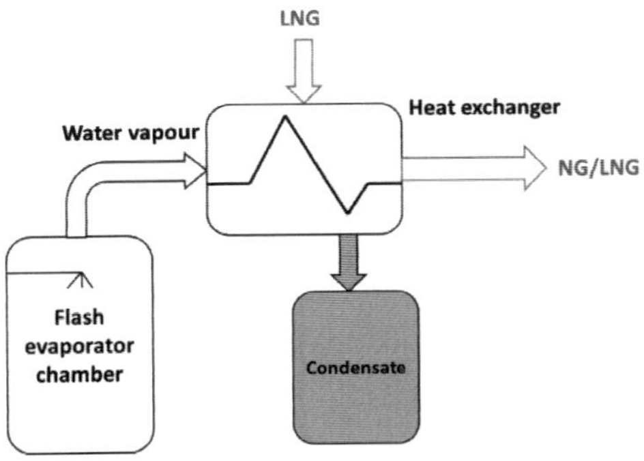

Figure 3. Block diagram of SCD-LNG system.

In order to evaluate the performance of the LNG cryogenic energy, only the condensation process in SCD and AWG methods are simulated. For each component, the governing equations are derived based on the principles of heat balance, mass balance, heat transfer, mass transfer and empirical equations from references. The model is programmed with the Siemens LMS Imagine. Lab Amesim in the specified time interval. In the SCD-LNG systems, the following assumptions are proposed to apply:

- There is no entrainment of mist by the vapour, and the vapour is salt-free.
- The vacuum in the condenser can be maintained throughout the simulation.
- The water vapour is considered as an ideal gas in the condenser.
- The water vapour and minor quantity air existing in the system is considered as the moist air.
- There is no heat transfer loss in the condenser.

Similarly, the assumptions of the simulation model of the whole AWG-LNG system are:

- There is no entrainment of mist by the moist air, and the moist air is salt-free.
- The heat exchanger is completely sealed.
- There is no heat transfer loss in the heat exchanger.
- All processes take place at steady state.
- The fluid flow pressure drop due to friction is neglected.
- Thermodynamics equilibrium is applied in the system components.

According to the assumptions above, the two models could be simulated into the same model with different operating parameters. The heat exchanger consisting of arrays of plates with flat tubes is designed. Moist air flows across the plates where it exchanges heat with the cooling medium LNG flowing through the tubes. When the condensation occurs, the heat transfer can be classified into 3 ways: heat convection between the wall and the moist air; heat convection between the wall and the cooling medium; heat conduction between the inner tube and outer tube.

When the condensation occurs in the condenser, the water vapour will be condensed at the condensation temperature and the phase changed from vapour to the liquid. During this process, the heat energy provided consists of the latent heat of phase change and the sensible heat of temperature reduction. The heat flow rate between the air and the wall during this process is given as:

$$q_{c,\,ma}=h_{ma}\cdot A_{c,\,ma}\cdot\left(T_{ma,\,in}-T_{w}\right)+\dot{m}_{ma}\cdot hv\cdot(ah_{w}-ah_{in}) \quad (1)$$

where h_{ma} is the convective heat transfer coefficient of the moist air side, the value is determined by the Nusselt Number (Nu), the thermal conductivity of the moist air and the characteristic length of the exchange as follows:

$$h_{ma}=\frac{\lambda_{ma}\cdot Nu}{l} \quad (2)$$

The condensate generated in the condenser during this process is gauged based on the inlet absolute humidity and the outlet absolute humidity:

$$\dot{m}_{cond}=\dot{m}_{ma}\cdot(ah_{w}-ah_{in})=\dot{m}_{da}\cdot(ah_{in}-ah_{out}) \quad (3)$$

where $\dot{m}_{da}$ is the mass flow rate of the dry air in the moist air, the value will be adopted from the moist air mass flow rate and the inlet relative humidity:

$$\dot{m}_{da}=\frac{\dot{m}_{ma}}{1+ah_{in}} \quad (4)$$

The outlet temperature of the moist air after condensation is determined by the inlet conditions and the heat transfer between the moist air and the wall:

$$T_{ma,\,out}=T_{ma,\,in}+\frac{q}{q_{max}}(T_{w}-T_{ma,\,in}) \quad (5)$$

Similar to the outlet temperature of the moist air, the outlet absolute humidity after condensation is expressed as :

$$ah_{out}=ah_{in}+\frac{q}{q_{max}}(ah_{w}-ah_{in}) \quad (6)$$

where q_{max} is the maximum heat transfer between the wall and the moist air. It is expressed as:

$$q_{max}=\frac{\dot{m}_{ma}}{1+ah_{in}}\cdot\left((1+ah_{sat})\cdot\left(H_{ma,\,in}-H_{ma,w,sat}\right)+(ah_{ma,\,in}-ah_{w,\,sat})\cdot(H_{ma,\,in}-H_{lw,w,\,sat})\right) \quad (7)$$

where $H_{ma,w,sat}$ is the moist air specific enthalpy at wall saturation conditions and $H_{lw,w,sat}$ is the liquid water specific enthalpy at wall saturation conditions.

With the cooling medium LNG entering the channels inside the wall, the internal convective heat exchange occurs between the inside wall and the cooling medium. The heat flow rate between the cooling medium LNG and the wall is expressed as:

$$q_{c,\,co}=h_{co}\cdot A_{c,\,co}\cdot(T_{co}-T_{w,\,co}) \quad (8)$$

where h_{co} is the convective heat transfer coefficient between the LNG and the wall, T_{co} is the temperature of the LNG in the channels, $T_{w,\,co}$ is the inside wall temperature.

In addition, the heat exchange of conduction between the walls should be taken into consideration due to the temperature difference between both sides and the material of the wall. Therefore, the heat flow rate of conduction is calculated as:

$$q_{w}=\frac{\lambda_{w}A_{w}}{\delta_{w}}(T_{w,\,ma}-T_{w,\,co}) \quad (9)$$

where δ_w is the thickness of the wall and λ_w is the thermal conductivity of the material.

Water recovery rate (RR) represents the recovery ability of the desalination plant which is the ratio of recovery fresh water to the feed water. For the SCD-LNG process, RR could be decided as shown:

$$RR = \frac{\dot{m}_f}{\dot{m}_{ma}} \times 100\% \quad (10)$$

where $\dot{m}_f$ is the freshwater production and $\dot{m}_{ma}$ is the water vapour flow rate. Similarly, for AWG-LNG system, RR is presented with $\dot{m}_{ma}$ as moist air flow rate:

$$RR = \frac{\dot{m}_f}{\dot{m}_{ma}} \times 100\% \quad (11)$$

3. VALIDATION

The two models have to be validated by experimental data from the test facility in Wu and Vierow [8]. A channel heat exchanger is designed to be simulated as shown in Figure 4.

In this literature, the condensation of vapour mixtures in the horizontal tubes is experimented to test the effect of the heat transfer characteristics and the temperature gradient across the condenser tube. The comparison covered a few operating parameters covering the air mass flow rate, temperature and pressure. The temperatures of the air mass at different location of the tube are compared as shown in Figure 5. The errors in the comparison of the temperature of steam between the literature and the LMS model are from 0.379% to 7.77%. The good agreements between the models in the LMS and the experimental data validate the models' capability to predict the system performance.

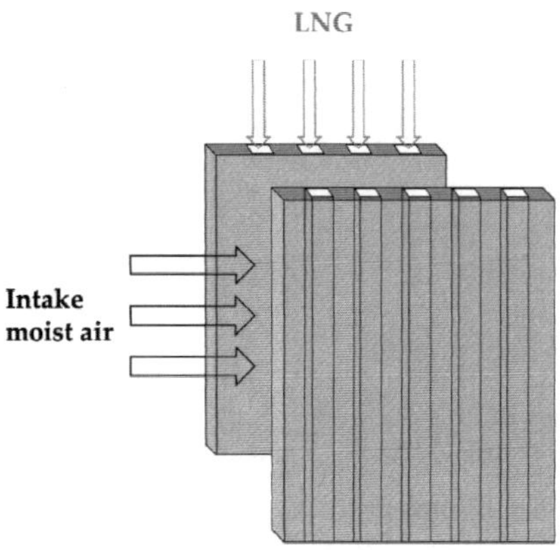

Figure 4. Schematic of the heat exchanger.

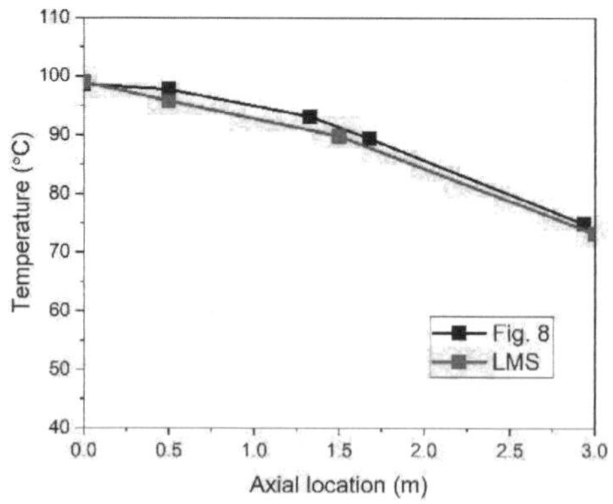

(a) Moist air parameters: pressure of 100 kPa, temperature of 99°C, air mass fraction of 15%, mass flow rate of 7 g/s.

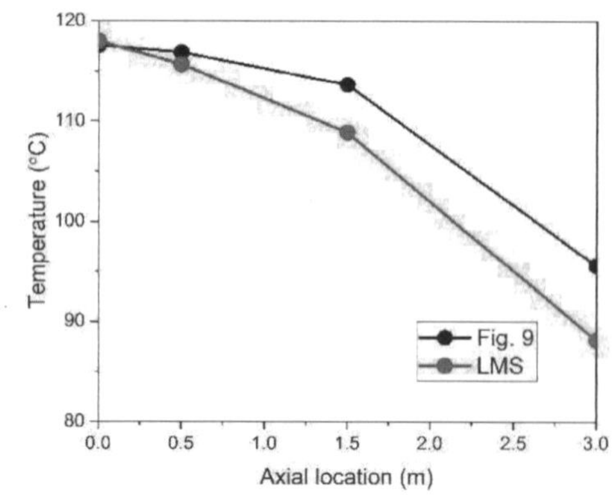

(b) Moist air parameters: pressure of 200 kPa, temperature of 118°C, air mass fraction of 15% and mass flow rate of 12 g/s.

Figure 5. Validation results comparison.

4. RESULTS

The validated model is now employed to evaluate the AWG-LNG system and SCD-LNG system under the different conditions as shown in Table 1 and Table 2. In order to simplify the simulation process, LNG is simulated as CH_4. The heat exchanger configuration parameters are decided: heat exchange area of 1.4 m^2, characteristic length of 3 mm, material of Al Alloy and thickness of 0.265mm. The performances of the two systems are discussed separately.

4.1 AWG-LNG desalination system

The climate data is obtained from Tariq, et al. [9]. The effect of the cooling medium and responding temperature are investigated accordingly. The performance of the proposed AWG-LNG system is compared with the same model cooled by cooling water of 5 °C as shown in Table 1. With the fixed moist air operating parameters and the configuration parameters of the heat exchanger, the recovery ratio (RR) of AWG-LNG is 41.8% higher than the normal cooling water model in Singapore climatic conditions. This means that when the LNG is applied as a cooling medium, the system efficiency has been promoted. And the consumed LNG cryogenic energy is 3 kW more than the cooling water. This phenomenon expressed that more LNG cryogenic energy is recovered to reduce the extra energy consumption.

Table 1. Performance of AWG-LNG comparison with water in Singapore climatic condition.

Parameters	Unit	Quantity
Moist air		
Temperature	°C	27.5±2.67
Relative humidity	%	84.0±4.7
Wind velocity	m/s	2.0±0.68
Cooling medium: LNG		
Temperature	°C	-60.0
Mass flow rate	kg/s	2.0
Pressure	Bar	1.0
Cryogenic energy consumption	kW	8.5
RR	%	1.91
Cooling medium: cooling water		
Temperature	°C	5.0
Mass flow rate	kg/s	2.0
Pressure	Bar	1.0
Cooling energy consumption	kW	5.1
RR	%	1.35

4.2 SCD-LNG desalination system

Similarly, as shown in Table 2, the performance of the proposed SCD-LNG system is employed for a cruise ferry with a service speed of 22 knots and utilizing a DF-Electric power plant. The average mass flow rate of LNG during the voyage at 0.875 kg/s is selected as the operation parameter. With the fixed moist air operating parameters and the configuration parameters of the heat exchanger, RR of SCD-LNG is twice higher than the normal cooling water model. And the LNG cryogenic energy has been recovered twice than cooling water system.

Table 2. Performance of SCD-LNG comparison with water.

Parameters	Unit	Quantity
Water vapour		
Pressure	Bar	0.08
Temperature	°C	41.5
Water vapour content	kg	2.485
Relative humidity	%	80
Mass flow rate	kg/s	0.5
Cooling medium: LNG		
Temperature	°C	-60
Mass flow rate	kg/s	0.875
Pressure	Bar	1.0
Cryogenic energy consumption	kW	155.4
RR	%	12.6
Cooling medium: water		
Temperature	°C	5
Mass flow rate	kg/s	0.875
Pressure	Bar	1.0
Cooling energy consumption	kW	80.7181
RR	%	6.49

In conclusions, the recovered LNG cryogenic energy provides a feasible way to reduce the cooling water supply and the costs for desalination. And the high temperature Although the ultra-low temperature of LNG as the cooling medium has shown the good cooling ability in the condensation process, the freezing phenomena may occur. Therefore, the research on preventing the ice formation on the cooling surface utilizing hydrophobic coating is developing.

5. CONCLUSIONS

The innovative LNG cryogenic energy assisted thermal distillation technology single-stage conventional distillation (SCD) and atmospheric water generator (AWG) is proposed in this paper. These two systems are modelled with Siemens LMS Imagine. Lab Amesim. The novelty of this work lies on the development of the condensation process assisted by the ultra-low temperature cooling medium: LNG. The validation models are first carried out under a series of experimental conditions in the literature. In order to explore the feasibility and the performance of the proposed SCD-LNG and AWG-LNG system, the freshwater yields are calculated and compared with the cooling water condensation systems. The following conclusions can be drawn from the discussions:

- The comparison results between the simulation in Siemens LMS Imagine. Lab Amesim and the experimental data in the literature presented that the model is capable to predict the performance of the proposed systems.
- The proposed SCD-LNG system and AWG-LNG system provides a reasonable alternative freshwater resource on LNG fuelled ships.
- With the fixed operation parameters of moist air and the configuration parameters of the heat exchanger, the lower temperature LNG as the cooling medium has presented a significant influence on the performance as predicted.
- The simulation result in SCD-LNG system shows that the efficiency has been promoted from 6.49% to 12.6% compared with the conventional cooling water system. And the AWG-LNG desalination system investigates that the efficiency has been promoted by 41.8%.
- The recovery amount of LNG cryogenic energy is maximum twice than the conventional cooling water system which has shown the potential of the proposed integrated system on board.

6. REFERENCE

[1] EIA. (2015). *Average operating heat rate for selected energy sources*. Available: https://www.eia.gov/electricity/annual/html/epa_08_01.html

[2] C. W. Graugaard, "LNG for Shipping – Current status," 2013.

[3] L. Register, "LNG-fuelled deep sea shipping," *Lloyds Register, August,* 2012.
[4] E. S. Jeong, "Optimization of power generating thermoelectric modules utilizing LNG cold energy," *Cryogenics,* vol. 88, pp. 29-35, 2017.
[5] B. B. Kanbur, L. Xiang, S. Dubey, F. H. Choo, and F. Duan, "Cold utilization systems of LNG: A review," *Renewable and Sustainable Energy Reviews,* vol. 79, pp. 1171-1188, 2017.
[6] W. Sun, P. Hu, Z. Chen, and L. Jia, "Performance of cryogenic thermoelectric generators in LNG cold energy utilization," *Energy Conversion and Management,* vol. 46, no. 5, pp. 789-796, 2005.
[7] R. S. Kumar, A. Mani, and S. Kumaraswamy, "Experimental studies on desalination system for ocean thermal energy utilisation," *Desalination,* vol. 207, no. 1-3, pp. 1-8, 2007.
[8] T. Wu and K. Vierow, "Local heat transfer measurements of steam/air mixtures in horizontal condenser tubes," *International journal of heat and mass transfer,* vol. 49, no. 15-16, pp. 2491-2501, 2006.
[9] R. Tariq, N. A. Sheikh, J. Xamán, and A. Bassam, "An innovative air saturator for humidification-dehumidification desalination application," *Applied Energy,* vol. 228, pp. 789-807, 2018.

7. AUTHORS BIOGRAPHY

Meixi Liu is a PhD student of Marine Technology at Newcastle University.

A PERFORMANCE AND EMISSION INVESTIGATION OF A MARINE DIESEL FUELLED WITH B20, B50 AND B100 MICROALGAE (SCHIZOCHYTRIUM SP.) OIL WITHOUT CHEMICAL ADDITIVE

H. M. Attar, Newcastle University, UK and King Abdulaziz University, SA
D. Wu and A. P. Harvey, Newcastle University, UK

SUMMARY

This study investigates the performance and emissions of a marine diesel engine fuelled with a raw microalga (Schizochytrium sp.) oil. The engine was fuelled with B20 (20% raw algae, 80% diesel), B50 (50% raw algae, 50% diesel) and B100 (100% raw algae oil) without chemical additives. The results were compared with diesel oil (DO), and at full load, the brake power of algae oil and its blends decreased by 26%. However, at lower loads, similar brake power results were observed for all test fuels. The BTE of algae oil has a variation of -22% and +10% when compared to DO. The algae oil results showed an increase in BSFC of 38%. The most critical finding from this research study and analysis is the significant reduction in NOx and CO emission values for the algae oil, in most test conditions. Therefore, these findings demonstrate that algae oil is the optimum alternative fuel for diesel engines and a promising solution to diesel engine emissions.

NOMENCLATURE

BSFC	Brake specific fuel consumption
BTE	Brake thermal efficiency
B20	20% raw algae, 80% diesel oil
B50	50% raw algae, 50% diesel oil
B100	100% raw algae oil
CO	Carbon monoxide
DO	Diesel oil
NO_x	Nitrogen oxides
RAO	Raw algae oil

1. INTRODUCTION

Energy consumption is the prime mover of economic development which has become a central issue for society. Moreover, the cumulative increase in world population coupled with economic activity has led to increasing energy demand. Scientists are aiming to find new sources of renewable energy instead of conventional energy to avoid petroleum crises and reduce pollution from exhaust gases [1]. The emissions generated from the electricity generation and transport sectors produce nearly two-thirds of global CO2 emissions, which significantly affects the environment. Biomass feedstock is a significant potential source of bioenergy. It plays a pivotal role in fossil fuel replacement because it is an effective strategy that can meet the global energy demand and reduce air pollution from fossil fuels [2].

Currently, biofuels are one of the most significant global contributors to renewable energy sources. Several methods and techniques can be used for converting biomass into bioenergy products, for example, photosynthesis, transesterification, gasification, fermentation and direct combustion [3]. Macroalgae oil is one of the potential alternative third-generation biofuels, particularly for internal combustion engines. Compared to conventional biofuel plants, it has a higher growth rate and lipids production up to 80% of dry weight [4]. Many researchers [5,6] have explored the advantages of algae biodiesels and demonstrated their potential as an alternative to fossil fuels. However, most of them focused on the production of algal fuel and the life cycle assessment, while the specific analysis for the combustion and emissions in diesel engines was rarely reported.

In general, microalgae oil is a non-edible biomass resource which can be used as an alternative fuel for IC diesel engines. Expensive chemicals and procedures associated with the transesterification reaction are the main problems related to the production of algae biodiesel. In this paper, the algae oil will be used as raw oil in order to overcome these problems. This research paper intends to investigate performance and emissions of the raw algae oil (RAO) and its blends on a marine diesel engine, without the addition of chemical additives, in terms of brake power, brake specific fuel consumption (BSFC), brake thermal efficiency (BTE), Carbon monoxide (CO) and Nitrogen oxides (NO_x).

2. MATERIALS AND METHODS

2.2 *MICROALGAE OIL*

A raw microalga oil extracted from Schizochytrium sp. biomass was used in the diesel engine without chemical additive. The properties of the RAO and diesel oil (DO) are illustrated in Table 1. The preheating process was applied to decrease the viscosity of the microalgae oil in order to alleviate any injection and combustion difficulties. The blending process for B20 and B50 was conducted by heating the RAO at 55-65 C° and blending it with pure diesel oil in a standard heating hotplate fitted with a blender mixer machine. The electric blender speed was set at 400 rpm for 30 to 40 minutes. For B100, the preheating of RAO was the only process used to prepare the samples before fuelling the engine. In the engine fuel system, the fuel passes through two other heating processes to maintain its temperature between 55 and 65 C°.

Table 1 Diesel oil and raw microalgae oil properties

Property	Unites	D.O	RAO
Kinematic viscosity at 50C°	cSt	1.811	19.67
Density	kg/m^3	762.5	930
Calorific value	MJ/kg	42.8	36.84
Carbon	%	86	75.9
Hydrogen	%	14	10
Oxygen	%	0	14
Flashpoint	C°	~52	180

2.3 *ENGINE SETUP AND TEST PROCEDURE*

The experiment test rig consisted of a four-stroke single-cylinder naturally aspirated compression ignition marine diesel engine connected to a 3-phase Prestolite alternator and electrical heater as a load bank. Engine specifications are listed in Table 2. The alternator was connected via toothed pulleys and a toothed belt. The electric output power of the alternator was dissipated in an air-cooled load bank mounted within the test rig frame.

A fuel flow measuring burette was connected to the fuel line via a two-way valve to allow fuels to be drawn from the burette when it switched to the measurement position. The microalgae oil tank is partially covered by a 180 W, 120 V flexible heating blanket and a 125W. 240V rope heater employed on a copper pipe before the fuel pump. The reason for this arrangement is to maintain the oil temperature during the experiments due to the viscous nature of the algae oil. Both heaters are connected to a PID temperature controller along with in order to monitor the fuel temperature. A TESTO 350 XL exhaust gas analyser is mounted on the exhaust manifold to measure both CO and NOx emissions. A schematic diagram of the engine test rig is shown in Fig.1

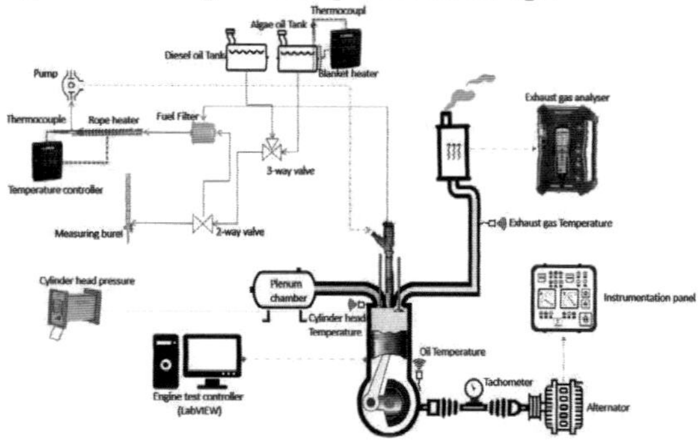

Figure [1] Schematic diagram of the engine test rig

The tests were carried out at different loads and speeds of 25%, 50%, 75% and 100% and 2000, 2400, 2800 and 3200 rpm. Initially, the baseline of the results was generated by running the engine with diesel oil. Before utilising the B20, B50 and B100, the engine was warmed up by diesel oil to intensify the average temperature within the combustion chamber (>500°C).

Table 2 Engine specifications

Lombardini 15 LD 225 Marine Diesel Engine	
Number of cylinders	1
Cycle	4-stroke
Power kW (HP)	3.5 (4.8)
Bore (mm)	69
Stroke (mm)	60
Fuel Consumption (gr/kwh)	267
Compression ratio	21:1
Injection system	Direct

3. RESULTS AND DISCUSSIONS

3.1 *BRAKE POWER*

The variation of brake power with engine speeds and loads is presented in Figure 2. The generated brake power of B20, B50 and B100 were almost the same when as diesel oil at 25%, 50%, and 75% load. At full load, the brake power of B20, B50 and B100 decreased by 32%, 39% and 26%, respectively. This may be as a result of the difference in the properties of algae and diesel oil such as lower heating value, viscosity, density and oxygen content. The higher viscosity of the algae oil can contribute to reducing the fuel pump efficiency, which enhances engine power. A reduction in brake power was reported when the microalgae biodiesel was tested in four-stroke diesel engines. The authors associated the reduction in the break power to the lower heating value, higher density and viscosity [7,8,9].

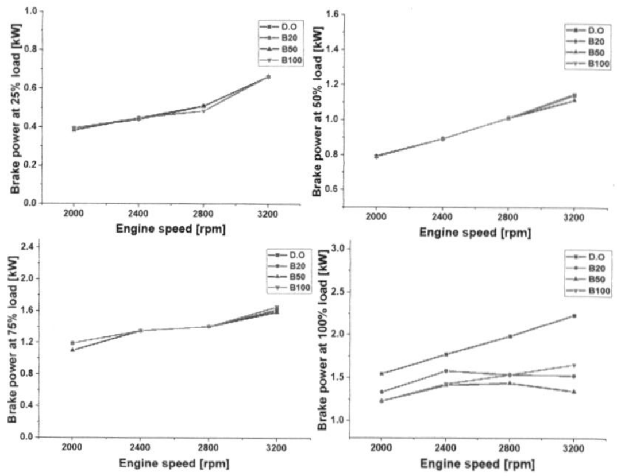

Figure 2: Brake power vs engine speed

3.2 *BRAKE SPECIFIC FUEL CONSUMPTION*

The brake specific fuel consumption is defined as a measure of the fuel flow rate per output power. The BSFC can determine how efficiently the fuel is consumed to produce the required work. The BSFC for all test fuels is illustrated in Figure 3. No significant differences were found in BSFC between B100 and DO. However, for B20 and B50, the engine consumed up to 22% and 38% more fuel when compared to DO. A possible reason for the increasing trend in BSFC is as a result of the lower heating value of the algae oil which forces the fuel system to inject more fuel in order to maintain the same speed [7,8,10]. Commonly, the high density and viscosity of the biofuels can lead to a decrease in fuel atomization, which causes an increase in BSFC [11].

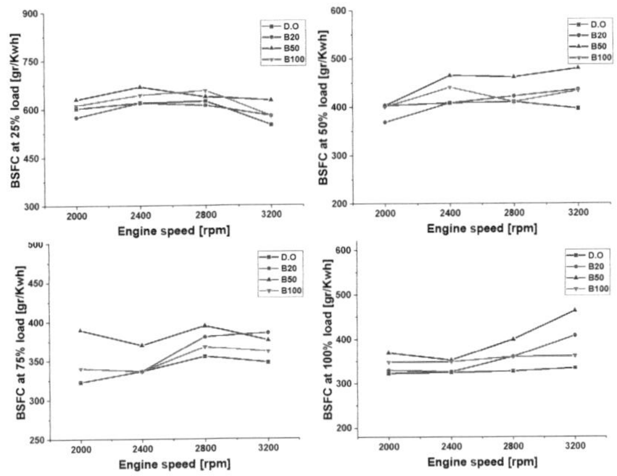

Figure 3: BSFC vs engine speed

3.3 *BREAK THERMAL EFFICIENCY*

Brake thermal efficiency is defined as the ratio of the brake power developed by the engine to the thermal input from the supplied fuel. Figure 4 illustrates the variation in the BTE for all test fuels. The BTE of B50 were lower than DO, B20 and B100 in all test conditions. This finding was also reported by Satputaley et al. [10] when a four-stroke diesel engine was fueled with microalgae biodiesel. The authors interpreted the reduction in BTE to the higher viscosity of the algae oil, which lead to poor fuel atomization. For B20 at lower speeds and loads, no significant differences were found compared to DO. However, at higher speeds and loads the BTE of B20 and B100 were 15% and 20% higher than DO in all test conditions. A possible reason for the increased in the BTE is as a result of the lower heating value of the algae oil compared to diesel oil [12].

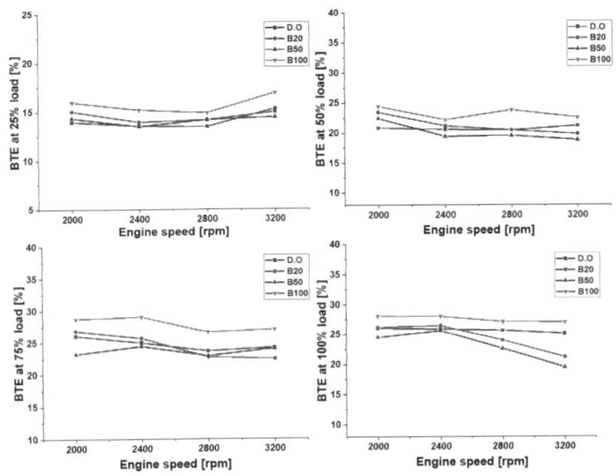

Figure 4: BTE vs engine speed

3.4 *CARBON MONOXIDE*

Carbon monoxide is a toxic flammable gas formed due to deficiency of oxygen during the combustion. Figure 5 presents the CO emissions of all test fuels. At full engine load, there is a clear trend of decreasing CO emissions, by up to 60%, for B100 at all speeds. The ignition delay of biofuel is longer than that of diesel, which will lead to a larger portion of fuel being burned during the premixed combustion phase. And this contributes to better fuel oxidation, and therefore less CO emission for algae oil and its blends. Satputaley et al. and Işık et al. [10,13] associated the reduction of CO emissions with the higher oxygen content in the algae oil, which leads to complete combustion. Conversely, DO results were better than B20, B50 and B100 at 25%, 50% and 75% load. This increased could be a result of the longer injection penetration of algae oil and its blends than that of DO, which will lead to fuel impingement on cylinder liner and higher hydrocarbon/CO emission.

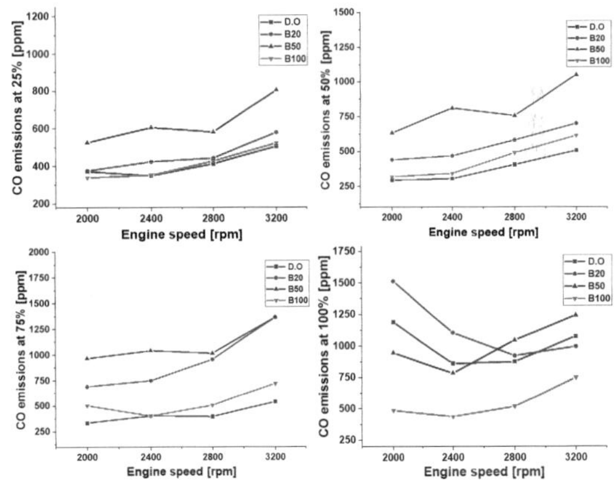

Figure 5: CO emissions vs engine speed

3.5 NITROGEN OXIDES

Nitrogen oxides (NOx) is one of the most critical emissions that diesel engines produce. The formation of NOx emissions is highly dependent on fuel oxygen concentration and combustion temperature. Figure 6 shows the NOx emissions variation for all test fuels. There was a dramatic decrease, by 59%, in NOx in all test conditions for B20, B50 and B100 load. Tüccar et al. [8] reported that the lower heating value of the microalgae oil affects the heat release rate and peak temperature, which explains this reduction in NO_x emissions.

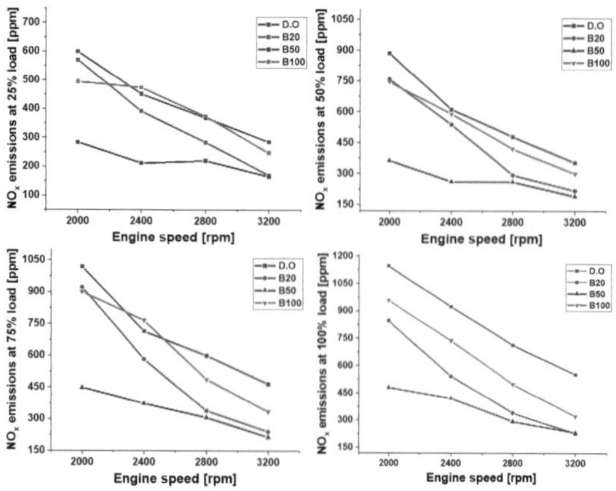

Figure 7: NO_x vs engine speed

4. CONCLUSIONS

Generally, the use of RAO and its blends as an alternative fuel can offer several environmental advantages. The effects of the RAO and its blends on brake power, BSFC, BTE, NO_x and CO emissions were investigated. A drop in brake power of up to 26% was observed at full load only, which is a result of the lower heating value, higher density and viscosity of the RAO. The lower heating value forced the engine to consume more fuel in order to produce constant output power which led to an increased in the BSFC by 38%. Also, the lower heating value affects the heat release rate and peak temperature, which reduced the NO_x emissions and improved the BTE. The longer ignition delay at full load leads to better fuel oxidation, and therefore less CO emission.

5. ACKNOWLEDGEMENTS

The authors wish to express their thanks to the Faculty of Maritime Studies, King Abdulaziz University, Saudi Arabia, for funding this study.

6. REFERENCES

1. Bahadar, A. and M. Bilal Khan, Progress in energy from microalgae: A review. Renewable and Sustainable Energy Reviews, 2013. 27: p. 128-148.

2. Kheshgi, H.S., R.C. Prince, and G. Marland, The potential of biomass fuels in the context of global climate change: Focus on transportation fuels. Annual Review of Energy and the Environment, 2000. 25: p. 199-244.

3. Bioenergy, I.E.A., Bioenergy–a sustainable and reliable energy source. International Energy Agency Bioenergy, Paris, France, 2009.

4. P. M. Schenk et al., "Second generation biofuels: high-efficiency microalgae for biodiesel production," vol. 1, no. 1, pp. 20-43, 2008.

5. Demirbas, M. F. J. E. c. Demirbas, and management, "Importance of algae oil as a source of biodiesel," vol. 52, no. 1, pp. 163-170, 2011.

6. D. L. Sills et al., "Quantitative uncertainty analysis of life cycle assessment for algal biofuel production," vol. 47, no. 2, pp. 687-694, 2012.

7. Al-lwayzy, S., T.F. Yusaf, and R. Al-juboori, Biofuels from the Fresh Water Microalgae Chlorella vulgaris (FWM-CV) for Diesel Engines. Energies, 2014. 7: p. 1829-18513390.

8. Tüccar, G. and K. Aydın, Evaluation of methyl ester of microalgae oil as fuel in a diesel engine. Fuel, 2013. 112: p. 203-207.

9. Sushant. S Satputaley, Chetan Chawane and N V Deshpande. Article: A Critical Review of Biofuels from Algae for Sustainable Development. IJCA Proceedings on National Conference on Innovative Paradigms in Engineering and Technology (NCIPET 2012) ncipet(9):7-11, March 2012. Full text available

10. Satputaley, S.S., D.B. Zodpe, and N.V. Deshpande, Performance, combustion and emission study on CI engine using microalgae oil and microalgae oil methyl esters. Journal of the Energy Institute, 2017. 90(4): p. 513-521.

11. Tsaousis, P., et al., Algae to Energy: Engine Performance Using Raw Algal Oil. Energy Procedia, 2014. 61: p. 656-659.

12. Al-lwayzy, S.H. and T. Yusaf, Diesel engine performance and exhaust gas emissions using Microalgae Chlorella protothecoides biodiesel. Renewable Energy, 2017. 101: p. 690-701.

13. Işık, M.Z., et al., The effect of n-butanol additive on low load combustion, performance and emissions of biodiesel-diesel blend in a heavy duty diesel power generator. Journal of the Energy Institute, 2017. 90(2): p. 174-184.

7. AUTHORS BIOGRAPHY

Hassan Attar, PhD student at Newcastle University. Teaching assistant at King Abdualaziz University.
Dawei Wu, EPSRC Research and Innovation Fellow, Lecturer in Marine Engineering
School of Engineering, Newcastle University
Adam P Harvey, Professor of Process Intensification & Head of Process Intensification Group, School of Engineering, Newcastle University

A COMPARISON OF HYDROGEN AND AMMONIA FOR FUTURE LONG DISTANCE SHIPPING FUELS

C. J. McKinlay, S. R. Turnock, D. A. Hudson, University of Southampton, UK

SUMMARY

Decarbonisation of the shipping industry is necessary. Two potential low-emission fuels for long distance international shipping are hydrogen (H_2) and ammonia (NH_3). Using data from an LNG tanker, approximations were made for energy requirements based on delivered power, with the maximum consumption for a single voyage being 9270 MWh. Calculations were made for the required volume, mass and variable cost for several fuel types. Results showed that H_2 required volume was 6550 m^3 and 11040 m^3 for liquid and pressurised gas storage respectively. H_2 is frequently dismissed for mobile applications due to low volumetric density, however these volumes are not unrealistic. Ammonia has several desirable characteristics, but also has low gravitational energy density increasing the overall ship mass by 0.3% to 3.7%, negatively affecting performance. Batteries are too large, heavy and expensive for long distance applications. Both hydrogen and ammonia have potential, but require further research before becoming viable.

NOMENCLATURE

AFC	Alkaline Fuel Cell
CCS	Carbon Capture Storage
ECA	Emission Controlled Area
EV	Electric Vehicle
EGR	Exhaust Gas Recirculation
FC	Fuel Cell
GHG	Greenhouse Gas
HFO	Heavy Fuel Oil
HHV	Higher Heating Value
IMO	International Maritime Organisation
LNG	Liquid Natural Gas
MDO	Marine Diesel Oil
PEM	Proton Exchange Membrane (fuel cell)
SCR	Selective Catalytic Reduction
CO_2	Carbon Dioxide
H_2	Hydrogen
NOx	Nitrogen Oxides
NH_3	Ammonia
SOx	Sulphur Oxides

1. INTRODUCTION

Climate change is increasingly becoming a major global concern and the reduction of greenhouse gas (GHG) emissions is imperative. International shipping accounts for 2.4% [1] of global emissions and, due to the primary fuel source being heavy fuel oil, also releases large levels of pollutants such as sulphur oxides.

The International Maritime Organisation (IMO) has set regulations with regards to pollutant emissions, namely nitrogen oxides (NOx) and sulphur oxides (SOx), from shipping by 2020, with the possibility of future further restrictions [1]. Furthermore, several nations and organisations have self-imposed stricter targets, for example the UK government's Clean Maritime Plan aims for zero emission shipping to be "commonplace globally" by 2050 [2].

Several different fuels have been outlined as potential future replacements for GHGs, with two of the most prominently mentioned being hydrogen (H_2) and ammonia (NH_3). However, other potential technologies include: biofuels, nuclear and Carbon Capture Storage (CCS) [3]. In addition, there are some working examples where ships operate using batteries and Liquid Natural Gas (LNG). It is important to understand the opportunities and challenges involved with these various fuel types, particularly for hydrogen and ammonia.

1.1 AIMS AND OBJECTIVES

The overarching aim of this project was to analyse likelihood of either hydrogen or ammonia being the primary fuel source for long voyage ships in the future, with a particular focus on the engineering challenges involved. To achieve this, the following objectives were outlined:

- review literature concerning hydrogen and ammonia as fuels to evaluate the critical challenges;
- gather data for a typical long-distance voyage such that it can be used to model different scenarios for hydrogen and ammonia solutions;
- analyse the engineering considerations (such as sizing) for various types of fuel cells or combustion devices; and
- identify barriers for hydrogen or ammonia use for shipping (e.g. safety regulations or supply).

2. CURRENT SHIPPING FUELS

2.1 OIL-BASED FUELS

For long distance international shipping, heavy fuel oil (HFO) is currently the most common fuel source [1]. HFO is a by-product of the crude oil refinery process and is consequently more economical for long distance applications than other oil-based fuels such as marine diesel oil (MDO). Other advantages include a particularly high energy density, thus requiring less space and weight than other fuel types. The combustion technology is generally mature, and therefore efficient and cost effective. Also, the storage process is simple and large quantities can be stored for long periods of time without depletion. This can lead to further economic benefits as the commodity is globally available and can therefore be bought in bulk in regions with lower prices.

However, HFO has a significantly higher sulphur content than diesel. This results in particularly high levels of the pollutant SOx when burning HFO. In addition, in 2012 international HFO ships produced 796 million tonnes of CO_2 [1] and therefore, for zero emission shipping to be achieved then many suggest that HFO is not a sustainable option.

2.1 (a) Scrubbers

Large vessels have a relatively long life cycle (around 25 years [4]). Therefore, several operators have researched methods of reducing emissions from current fleets. For example, fitting a post-combustion marine gas exhaust system or 'scrubber' can reduce SOx levels. Other options include Marine Diesel Oil (MDO) which contains less than 0.1% sulphur [4], or 'desulphurised' HFO [5].

Currently, scrubbers are the most economical for long distance applications [5] and can meet the IMO's regulations. However, usage increases the overall energy demand of the vessel [1], raising the amount of GHGs being burned. Consequently, scrubbers tend to actually increase CO_2 emissions [5]. Therefore, the development of post-combustion techniques is unlikely to significantly increase the sustainability of HFO as a marine fuel, unless there is a major advancement in CCS.

2.2 LIQUID NATURAL GAS (LNG)

Natural gas is a commodity with many possible applications, for example it is the primary source of domestic heating in the UK [6]. The vast majority of oil reservoirs contain natural gas, as the global demand for oil is high, the supply of gas is relatively large and is consequently relatively inexpensive (around 2.2 p/kWh [7]). Over time, as the price of natural gas reduced, then larger and more frequent volumes began to be shipped.

Other advantages of natural gas include existing safety protocols [8] and, due to a higher hydrogen-to-carbon ratio [9], significantly fewer emissions and pollutants than HFO. However, the energy density is lower than HFO and the storage is notably more complex.

To reduce the volume required for storage, the majority of natural gas is transported as a liquid, reducing volume requirements by 600 times [10]. Given that the boiling point of natural gas is around -160°C [8], then LNG tankers require relatively advanced technology in the form of: insulated storage tanks with 'metallurgical' properties, and specialised double-hull design [10].

When storing large quantities in liquid form, there will be a certain level of 'boil off' [11]. Originally, this was simply lost fuel, so to capitalise on this situation many marine manufacturers designed LNG tankers to be powered by a steam turbine. This meant that the tanker could use either oil or LNG.

Gradually LNG tankers became more economically viable, evidenced by the global number of LNG tankers increasing from around 60 in 1990 to over 300 by 2008 [10]. Furthermore, global trade of LNG rose from 100 million tonnes in 2000 to 319 million tonnes in 2018 [12]. This is relevant for alternative fuels, as it demonstrates that once a type of tanker passes below a certain price point then global fleets can grow at a fast rate.

Furthermore, there is a direct correlation between the growth of LNG as a fuel and its high shipping demand. Therefore, large projected imports for a fuel (e.g. hydrogen) could indicate a higher probability of it powering ships.

Some sources suggest that a blend of natural gas and hydrogen may be the most economic option for future transport fuel [13]. However, the use of natural gas in any capacity will still produce a certain amount of GHG emissions unless there are significant developments in the field of CCS. Therefore, this approach would likely be considered a short term solution, especially when targeting zero emission shipping.

3. HYDROGEN

Hydrogen in its pure form (H_2) could be a solution for future sustainable marine transport, either by combustion or using fuel cells [14] [15] A few hydrogen-powered ships have been developed, but all to date have been designed for relatively short distances [16].

3.1 EMISSIONS

Hydrogen has no carbon content, and therefore has the potential to be emission free at the point of use, for example the only by-product of a Proton Exchange Membrane (PEM) fuel cell is water. However, some sources suggest that burning hydrogen with air (as opposed to pure oxygen) at certain temperatures can lead to the release of the pollutant NOx [17]. Nonetheless, there would be significantly lower levels of pollutants than other fuel options (notably ammonia).

However, hydrogen should only be considered truly carbon neutral when the production is emission free [18].

3.2 SUPPLY

Security of supply has been highlighted as a potential barrier to the viability of hydrogen [18]. Hydrogen is the most abundant element on the planet, but is typically found in the more stable form of water, with less than 1% being readily available as a gas [17].

Currently around 50 billion kg per annum [19] of hydrogen is produced and consumed worldwide this would be enough to supply around 250 million vehicles for a year (based on a consumption rate of 60 kg/mile

[19] and an average annual mileage of 12000). The majority of this is produced using a technique called steam reforming, this uses fossil fossils such as methane or natural gas and is therefore quite carbon intensive. It is, however, quite a mature technique and has a relatively high efficiency.

It is possible to produce emission free hydrogen using electrolysis with electricity from renewable sources. Electrolysis technology has developed considerably in recent years, with several sources suggesting that the PEM electrolyser should now be considered commercially viable [20] [16].

Currently the main global application of hydrogen is for the production of ammonia, accounting for around 53% (26.5 billion kg) of total usage [19]. This ammonia is mainly used for fertilisation. Therefore, production and distribution of hydrogen would have to increase, should either hydrogen or ammonia become a globally used fuel.

Hydrogen is also used in the oil refinery process, with 3.8 billion kg being used by the industry every year in the USA [19]. Therefore, should hydrogen become a more prominently used fuel, then it is almost inevitable that the demand for oil would fall. Consequently, the hydrogen that would typically be used for oil refinery could then help to meet the increased hydrogen demand.

3.3 REFUELLING

To be financially viable, it is likely that any future shipping fuel will require a relatively short period of time for refuelling, ideally taking less time than the cargo loading/unloading process. Fast turnaround is important in shipping as it increases a vessel's maximum number of voyages per year, resulting in a "higher revenue earning potential" [21]. It is anticipated that refuelling time for hydrogen would be comparable to diesel [22]. Although further research and development of the refuelling process and infrastructure may be required.

3.4 PERFORMANCE

There are several possible methods for converting chemical potential energy to kinetic energy. For example steam engines boil water to power a turbine. More commonly for diesel, an internal combustion engine is use (ICE) which induces a series of explosions within cylinders that are positioned such that this will then rotate a driveshaft [23].

ICEs are not exclusive to oil-based fuel and the technology exists such that either hydrogen [24] or ammonia [25] could be used. However, due to the different higher heating values (HHV) of fuel types then ICEs tend not to be very versatile in terms of the feedstock. Therefore, to use these alternative fuels effectively, then specialist equipment would be required.

The advantage, however, of hydrogen's high HHV is that, coupled with its low gravitational energy density, hydrogen fuel may be, not only cleaner, but also more efficient than current fuels [26].

3.5 SAFETY

Hydrogen in its natural state is unscented, non-toxic and invisible [27], therefore leakage can be difficult to detect. This is a concern as H_2 is explosive, with a flammability range of between 4% and 77% when mixed with air [43].

Hydrogen would have to comply with EU regulations for all fuels, including specifications such as keeping containers in well-aired locations and away from ignition sources [22]. However, given the unique properties of hydrogen, then it is possible that specific global safety protocols would be deemed necessary. This would potentially restrict the growth of distribution and cause unforeseen expenses for hydrogen tanker operators.

3.6 STORAGE

A key engineering challenge for hydrogen fuel is storage. The natural state of hydrogen at ambient temperature is gaseous. It is both lighter than air and in a more stable state when in the form of H_2O. Therefore, it is anticipated that some form of closed tank will be necessary to store the hydrogen for marine purposes. Generally, the most common methods for storing hydrogen can be categorised into the following: pressurised gas, liquid, or metal hydride.

3.6 (a) Pressurised Gas

The critical temperature of H_2 is 33 K [28], meaning that it is simplest to store as a gas. However, at ambient temperature and a pressure of 1 bar, hydrogen has a volumetric density of 0.09 kg/m^3 (gasoline is 4.4 kg/m^3) thus requiring a significantly larger fuel tank [29].

To reduce volume requirements, hydrogen can be stored in a highly pressurised container, up to around 700 bar. Korri argues that a significant technological breakthrough is required before compressed hydrogen gas storage is financially viable, as volumetric density does not increase linearly with increasing pressures [13]. Additionally, the cost of carbon fibres (required for these pressures) are relatively expensive [13]. However, Korri's paper only provides an overview of the technology and does not consider the feasibility for specific applications (such as shipping).

An additional consideration for pressurised gas hydrogen is the casing requirements. For example, often steel casing is used. It may be necessary to consider the impact of the weight of the casing on performance.

3.6 (b) Liquid Hydrogen

Alternatively, hydrogen can be stored as a liquid to further increase the volumetric energy density, yielding the equivalent HHV/volume as pressurised gas at 800 bar [27] [53]. To store hydrogen as a liquid, temperatures between 13.8 K and 33.2 K are required [22]. Considerable energy is required to reach and sustain temperatures in this region, causing a projected increase in total energy demand up to 30% [27]. Also, due to more sophisticated technology being required, some suggest that capital costs could be 4 to 5 times higher than pressurised hydrogen gas storage [27].

Furthermore, the 'boil-off phenomenon' could be a drawback to this storage method [13]. However, LNG tankers harness and utilise the boil off, and therefore it may be feasible that hydrogen-powered ships could operate in a similar fashion. Alternatively, boil-off gas can be re-liquefied, however there is an energy cost to this process.

3.6 (c) Metal Hydrides

Metal hydride is a storage method where large amounts of hydrogen are absorbed into metals using chemical bonding [14]. In terms of volume, this is the most effective method of storing hydrogen.

3.7 COMBUSTION VS FUEL CELLS

Due to the high flammability of hydrogen, combustion is relatively uncomplicated, especially for steam turbines (such as used for LNG tankers) as they use different fuels interchangeably. ICEs, however, would require design modification due to the different HHV of hydrogen and gasoline [30].

The use of a fuel cell (FC), a device that uses the energy stored in chemical bonds to create electricity [31], is the most efficient method for extracting energy from hydrogen [48]. To power a ship, the FC would first produce electricity, then this would supply power to a motor. The use of an electric drivetrain would require fundamental design changes for ships, and therefore, it is probable that combustion of hydrogen would be more economical in the short term. Although, this may be considered a transitional phase as FCs are likely to be the most efficient and economical solution in the long term [32]. The PEM is generally considered the most economical and efficient FC type for pure hydrogen [33].

3.8 FUTURE CARGO PREDICTIONS

The growth of LNG as fuel can largely be attributed to the demand for LNG transportation, as utilising the cargo proved to be efficient and cost effective. Therefore, future demand projections for hydrogen, may be a strong indicator of the probability of it becoming a marine fuel.

For example, the Japanese government has invested large amounts into hydrogen development and infrastructure, with a total expenditure estimated to be over 12 billion USD (£9.6 bn) [31]. Consequently, sources suggest that Japan will have 800,000 hydrogen powered cars in the country by 2030 [31]. To meet the demand, the Japanese government projects that beyond 2030 annual imports of between £5.5bn and £11bn of hydrogen will be required. Furthermore, reports suggest South Korea and Singapore also anticipate "large levels" of hydrogen import [31]. Therefore, the likelihood of large quantities of hydrogen being shipped internationally in the near future is relatively high.

4. AMMONIA

Since the development of the Haber-Bosch process (a synthetic manufacturing technique) in the early 20th century, ammonia has been distributed globally, primarily as a fertiliser [34]. Due to its high hydrogen content, several sources suggest that ammonia could be a future fuel option for shipping [31] [2] [25] [32]. Similar to hydrogen, ammonia could either be burned or used in a fuel cell. Additionally, a third option would be to use ammonia as a hydrogen carrier.

4.1 EMISSIONS

Like hydrogen, ammonia (NH_3) has no carbon content. However, due to its high nitrogen content, the combustion of ammonia can release high levels of NOx. The release of NOx has numerous detrimental environmental impacts, such as: producing acid rain and smog, reducing air quality, and damaging the Ozone layer [35]. This was one of the main driving factors behind IMO bringing in strict regulations on maximum NOx emission levels. Therefore, it is fundamental to ammonia becoming a mainstream fuel that methods are developed to reduce this specific emission type. Some reports have suggested that early testing of post-combustion technology (such as catalytic converters) have been relatively promising [36], however it is unproven whether these techniques will be sufficient. Also, as with sulphur scrubbers, post-combustion processes can markedly increase energy consumption.

4.2 SUPPLY

A drawback of ammonia is the high energy intensity of its manufacturing process. Current production accounts for 2% of global energy consumption and 1% of CO_2 emissions [31]. For ammonia to become a major fuel type for international shipping, then production would be required to increase significantly, bringing uncertainty whether this demand could be met. Part of the reason for this high energy demand is because it requires hydrogen as a feedstock.

Logan discussed a technique for recovering ammonia from wastewater [19]. This is a resourceful approach, however given the minuscule ammonia content in water, it is unlikely to make a notable different to supply levels.

4.3 REFUELLING AND SAFETY

At 10 bar of pressure, ammonia is a liquid, therefore it is anticipated that the refuelling process for an ammonia powered ship would be a relatively straight forward and fast process. However, ammonia is severely corrosive, with a toxicity over three orders of magnitude higher than comparable fuels such as methanol or diesel [37] [38].

One of the advantages of ammonia having a pre-existing global supply chain, albeit mainly for the fertiliser market, is that there are already some globally universal safety protocols in place. However, as this is a new application, additional measures may be required, potentially leading to time delays and increased expenses.

One advantage of ammonia, in terms of safety, is it is not flammable in air [37] and is therefore less likely to cause fires and explosions than comparable fuels.

4.4 STORAGE

Ammonia can be stored as a liquid at ambient temperature by applying only 10 bar of pressure [38], or at ambient pressure with temperature of -10°C [31]. Thus making it significantly easier to store than hydrogen.

4.5 COMBUSTION VS. FUEL CELLS

As with hydrogen, a fuel cell is the most efficient method to extract energy from ammonia. Solid Oxide (SOFC) is considered the most efficient FC type for this fuel, as it can be directly fed with ammonia. This is opposed to PEM or Alkaline (AFC) that are fed exclusively from hydrogen, and therefore require ammonia cracking [55].

A further consideration for combustion is that ammonia is not flammable in air and therefore often requires a feed stock (such as hydrogen) to start ignition [37].

4.6 AMMONIA AS A HYDROGEN CARRIER

An alternative application would be using ammonia as a 'hydrogen carrier'. Here, ammonia would be stored and transported but converted back to hydrogen before being used for ship propulsion. There are several chemicals that could potentially be hydrogen carriers, but only ammonia has no carbon content [39].

An advantage, over using ammonia as a fuel, is the removal of NOx emissions, assuming that the conversion process from NH_3 to H_2 does not emit any.

The benefit over a purely hydrogen ship, is less complicated storage, as Service suggests that the "energy penalty of converting the hydrogen to ammonia and back is roughly the same as chilling hydrogen" [31]. This is due to the large energy requirement to maintain the low temperatures for liquid hydrogen, and also the high efficiency of the conversion process at around 99% (by heating ammonia to 200°C) [40].

One key consideration, however, would be the additional infrastructure required and more difficult logistics if performing this process on board a vessel.

4.7 FUTURE CARGO PREDICTIONS

One market intelligence firm predicts a growth rate in ammonia of 5% between 2019 and 2024. This is partially based on the potential market of ammonia in refrigeration, but also the hindrance of the hazardous effect of ammonia. By comparison, the same resource estimates a similar growth (5.4%) in hydrogen over the same period [41].

5. METHODOLOGY

To approximate the energy requirements of a long haul vessel, a dataset for a typical LNG tanker (LNG01) was analysed.

The building and testing of alternative fuel ships would entail large capital expenditure. Therefore, this research could cause significant economic benefits by providing indications of the feasibility of different future fuel types.

5.1 RESOURCES

The dataset for the LNG tanker contained several parameters logged at 5-minutely intervals, including: shaft power, fuel oil mass levels, longitudes and latitudes.

LNG01 has five cryogenic storage tanks, other key characteristics of LNG01 are shown in Table 1.

Table 1: LNG01 key details

Detail	Figure
Deadweight (summer)	73000 t
LNG storage tanks	5
Length	290 m
Breadth (maximum)	46 m
Total storage capacity	135000 m^3
Fuel Oil tank volume	2700 m^3
Main Engine	Mitsubishi marine steam turbine engine
Engine Output	Max. 21,320 kW x 81 rpm

5.2 ASSUMPTIONS

For a modelling exercise, a reasonable amount of assumptions are required. An assumption for this study was that the data collected for this ship is a reasonable representation for all long-haul shipping. The dataset has been recorded over a reasonably large number of voyages (108) and over a relatively long period of time (3 years 2 months) thus reducing the likelihood of anomalies influencing the accuracy of the data. Additionally, the operations and practices of LNG01 are understood to be typical for an LNG tanker.

International shipping is not just limited to tankers, and the intention of this study was that it could be applied to other categories of long distance shipping. Therefore, the assumption has been made that LNG tanker data is comparable for other ship types, particularly large cargo freighters. There exist examples of models where a similar assumption has been made [42].

Figure 1 shows a comparison between the standard cross-sections for tankers and 'light' cargo ships. These designs are not exactly the same, and the LNG01 design has additional design alterations (i.e. the cargo holds are spherical). However, it appears that the similarity is close enough to justify this assumption.

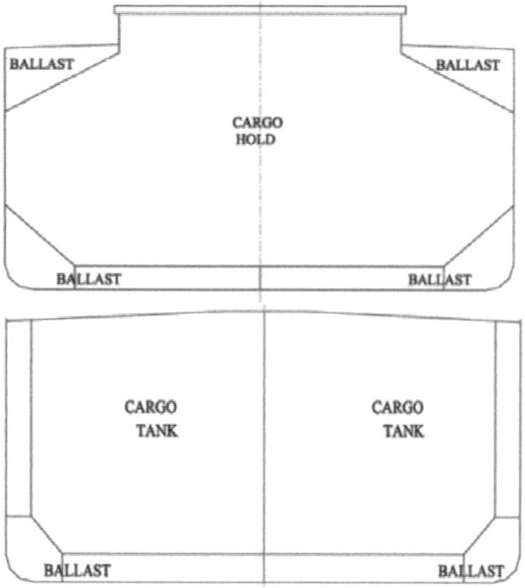

Figure 1: Cross-sections of a typical light cargo ship (left) and a typical tanker (right) [42]

The term 'light cargo' is used to distinguish from ore cargo that has very high density and therefore requires more ballast. Therefore, LNG01 would not be representative of an ore tanker.

As the majority of calculations have been based on the shaft power readings, this has implicitly accounted for the influence of weather and waves. Furthermore, the data was collected over a long time period for multiple voyages. Therefore, it is reasonable to assume the impact of weather would have evened itself out, such that this is a fair representation of a typical large vessel.

Nevertheless, care should be taken when considering individual cases or for specific parameters. For example, speed has been provided in two different readings: speed log (measured ship speed), and speed GPS (geographical speed). These readings generally follow the same trends, and are within 1 knot of each other for 77.2% of the data.

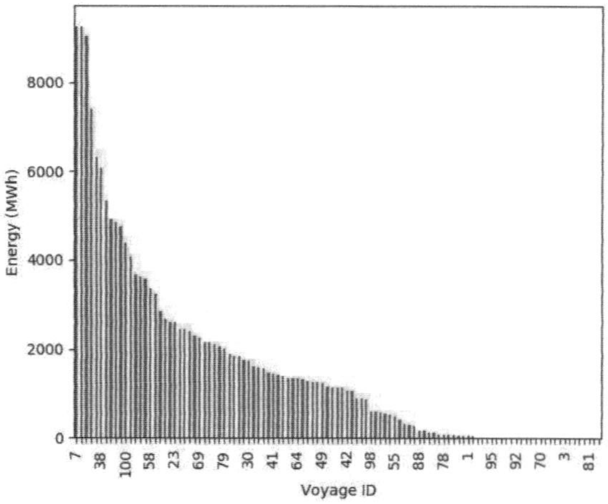

Figure 2: Total delivered energy (MWh) based on shaft power readings for each of the 108 voyages. Displayed in descending order.

This study has not accounted for the energy required for auxiliary power (on-board energy systems). Although it is anticipated that this would account for a small percentage of total consumption, it may be valuable to consider in future, especially as this power output may vary depending on the fuel type (e.g. additional energy for cooling/heating hydrogen).

6. RESULTS AND DISCUSSION

6.1 ENERGY CONSUMPTION

The dataset covered a period of 3 years 2 months, divided into 108 individual voyages (i.e. 108 opportunities for refuelling). Based on the reading of 'Shaft Power' at 5 minutely intervals, it was possible to make a reasonable estimate of the total delivered energy for each voyage, see Figure 2. This reading has been used throughout this study, as it is independent of the efficiency of the overall power train.

It is observable from Figures 2, 3 and 4, respectively, that journeys tend to vary considerably for energy consumption, distance and duration. The maximum energy requirement was for voyage number 7 at 9270 MWh, this will form the initial basis for required fuel storage calculations.

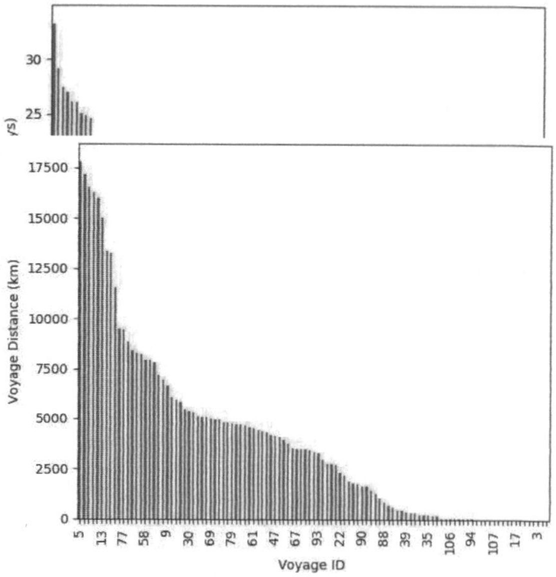

Figure 3: Total distance (km) for each of the 108 voyages. Displayed in descending order.

6.2 FUEL COMPARISON TABLE

Based on the findings in Section 6.1, it was possible to create Table 2 that compares requirements for volume, mass and cost (at current market price). This table was created initially based on the maximum expected delivered energy consumption for any given voyage: 9270 MWh.

Additionally specifications from the tanker's technical documents were used for: total storage capacity (135000 m^3); deadweight (63450 tonnes); fuel oil tank volume (2700 m^3) and weight capacity (2430 tonnes).

The calculations for Table 2 have been based on several estimates for the efficiency that the chemical potential energy could be transformed into propulsion. For example, the combustion of Hydrogen would have an efficiency of 40%, whereas fuel cells would be 55 to 60% [43]. It is noteworthy that the increased uptake of a technology tends to lead to further development, therefore the efficiencies for alternative fuels would likely increase. For each of the efficiency ranges in Table 2, the higher boundary value has been used.

6.2 (a) 6350 MWh Table

It is observable from Figure 2 that the majority of voyages have considerably lower energy demand than 9270 MWh. For example, had the maximum energy storage been 6350 MWh, then 96% of journeys would have been achievable. The risk of running out of fuel far from a port is a serious one but, with correct fuel management, it is deemed feasible that a long distance tanker could operate effectively under the latter capacity.

Table 3 compares the requirements of each fuel option for this lower energy requirement. It is observable that some of these figures may be more achievable, such as 4480 m^3 for liquid hydrogen storage.

6.3 HFO TANK COMPARISON

The majority of LNG tankers are primarily fuelled by LNG itself, but also contain an oil tank that can provide additional power. As this oil tank is purely for propulsion and not part of the cargo, then comparisons to this have been drawn. It is noteworthy that this fuel tank (highlighted in Figure 5) could be filled to 88% and still provide enough power for all historical voyages. This has provided fuel security, and the ability to purchase large quantities of fuel in cheaper regions.

This strategy is effective as oil has a particularly high volumetric energy density and is easy to store, with negligible losses when storing for a long period of time. However, other fuel options, particularly hydrogen, do not share these characteristics. Therefore, it is possible that the trade-off between the costs of increasing the storage tank versus the flexibility of purchasing, may not be cost effective. Nonetheless, if hydrogen were to become the fuel of choice for a large percentage of ships, then production and distribution would have to increase significantly and as a result it is particularly difficult to attempt to predict future economics for the fuel.

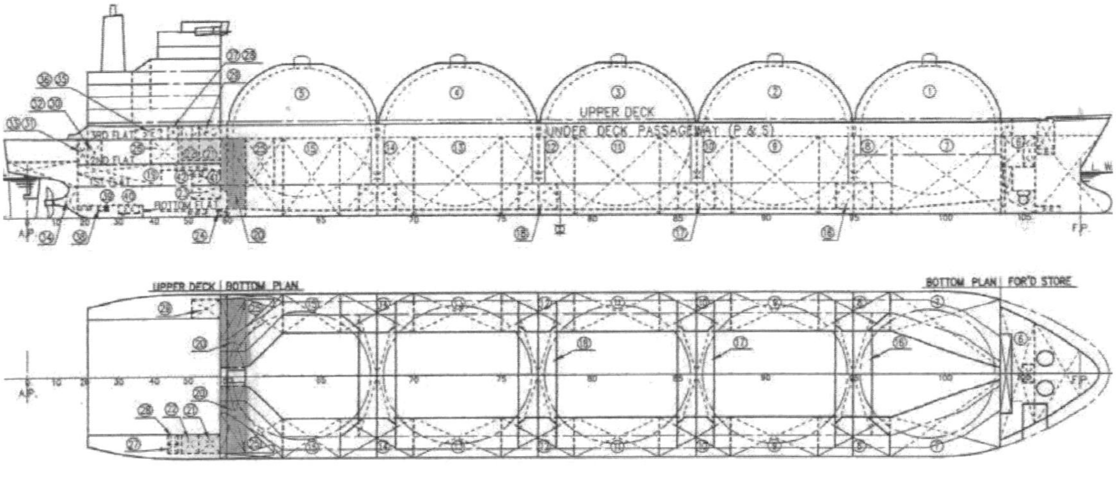

Figure 5: Size and location of the current fuel oil tank (highlighted right) and support tank (highlighted left)

Table 2: A comparison of fuel volume, mass and cost to provide 9270 MWh

Fuel type	**LNG** [44] [7] [45]	**Diesel (HFO)** [46] [47] [48] [49]	**Hydrogen (gas @ 700bar)** [46] [50] [51] [48]	**Hydrogen (liquid)** [46] [51] [48]	**Metal Hydride** [46] [51] [48]	**Ammonia** [52] [53] [54] [55]	**Methanol** [52] [53] [56]	**Batteries (Li-ion)** [57] [58] [59]
Efficiency	58%	20-40%	40-60%	40-60%	40-60%	30-60%	55-60%	70-95%
Required input energy (MWh)	15983	23175	15450	15450	15450	15450	15450	9758
Volume								
Energy density (MWh/m³)	5.83	9.7	1.4	2.36	3.18	4.82	4.99	0.30
Total storage size (m³)	2740	2389	11036	6547	4858	3206	3095	32855
% of cargo	2.03%	1.77%	8.17%	4.85%	3.60%	2.37%	2.29%	24.3%
% compared to max FO	101%	88%	409%	242%	180%	119%	115%	1217%
Mass								
Energy density (MWh/kg)	0.0142	0.0116	0.0333	0.0333	0.0006	0.0052	0.0055	0.0002
Total storage mass (tonnes)	1123	1998	464	464	26638	2959	2792	44354
% of total	1.68%	2.99%	0.69%	0.69%	39.81%	4.42%	4.17%	66.3%
Price								
Fuel per voyage (£ Millions)	0.349	1.367	8.654	8.654	8.654	1.976	1.123	6.913*

Table 3: A comparison of fuel volume, mass and cost to provide 6350 MWh

Fuel type	**LNG** [44] [7] [45]	**Diesel (HFO)** [46] [47] [48] [49]	**Hydrogen (gas @ 700bar)** [46] [50] [51] [48]	**Hydrogen (liquid)** [46] [51] [48]	**Metal Hydride** [46] [51] [48]	**Ammonia** [52] [53] [54] [55]	**Methanol** [52] [53] [56]	**Batteries (Li-ion)** [57] [58] [59]
Efficiency	58%	20-40%	40-60%	40-60%	40-60%	30-60%	55-60%	70-95%
Required input energy (MWh)	10948	15875	10583	10583	10583	10583	10583	6684
Volume								
Energy density (MWh/m³)	5.83	9.7	1.4	2.36	3.18	4.82	4.99	0.30
Total storage size (m³)	1877	1637	7560	4484	3328	2196	2120	22506
% of cargo	1.39%	1.21%	5.60%	3.32%	2.47%	1.63%	1.57%	16.67%
% compared to max FO	70%	61%	280%	166%	123%	81%	79%	834%
Mass								
Energy density (MWh/kg)	0.0142	0.0116	0.0333	0.0333	0.0006	0.0052	0.0055	0.0002
Total storage mass (tonnes)	770	1369	318	318	18247	2027	1913	30383
% of total	1.15%	2.05%	0.47%	0.47%	27.27%	3.03%	2.86%	45.40%
Price								
Fuel per voyage (£ Millions)	0.239	0.936	5.928	5.928	5.928	1.354	0.769	4.735*

*Based on a 20 year life cycle and an electricity price of 12.5 p/kWh

6.4 BATTERIES

It is observable from Table 2 that Lithium-ion batteries, despite being the most widely used type of battery for mobility applications (such as electric vehicles) [60], require significantly more space and weight. Furthermore, the initial capital cost for a battery capacity of 9758 MWh has been estimated to be £3.9 billion (502 $/kWh [23]). This has been factored into Table 2 on a per voyage basis, in addition to an estimated electricity cost.

Due a significant increase in global Li-ion production, it is projected that the price of the batteries will decrease, with 2030 price projected as 140 to 620 $/kWh [61]. However, lower value of $140/kWh would return a total cost of £1.04 billion which is still considerably higher than the alternatives. Additionally, recharging time is a concern, the Tesla supercharger has a power rating of 240 kW [62] therefore it would take 1000 superchargers 40.7 hours to recharge 9758 MWh, also the local distribution network may struggle to meet the additional demand.

Consequently, it has been concluded that batteries are not a viable option to provide the primary power supply for long distance shipping in the foreseeable future. Nevertheless, batteries could be useful for other marine applications, such as smaller short distance ferries or as part of a hybrid power management system, especially when using fuel cells [20].

6.5 HYDROGEN VOLUME REQUIREMENTS

One of the major criticisms of hydrogen as a fuel is the perceived low volumetric energy density, as shown in Table 2. However, in comparison to the fuel oil tank, this would only require a 242% increase in size using liquid storage. It is observable from Figure 5 that the fuel oil tank takes up a relatively small percentage (2%) of the vessel's capacity.

From Table 3, calculations showed that 6350 MWh could theoretically be provided by 2284 m^3 of liquid hydrogen, thus implying that increasing the current space allocated for the fuel oil tank (2700 m^3) by 66% would be sufficient. However, this considers only the fuel itself and further research is required into the sizing of the storage tank, which may be larger for cryogenic storage.

A metal hydride system requires significantly less volume than other hydrogen storage methods. Although, Table 2 shows the weight of this system (26.6 million kg) is several times the magnitude of most of the other fuel options. Therefore, it has been concluded that metal hydrides should not be considered a viable option for mobility applications in the near future.

6.6 HYDROGEN PRICING

The hydrogen columns in Table 2 show that the current price is considerably higher than some of the comparable fuels. This figure, £8.65m, is based on the current price per litre to refill a hydrogen vehicle at a gas station [51]. However, given that a vessel of this scale would be purchasing in large quantity, the commercial price would be expected to be lower.

Hydrogen production has historically tended to use processes (such as steam reforming) that use fossil fuels as the primary feedstock [63]. However, in recent years the efficiency of electrolysis has improved. Electrolysers are effectively reverse fuel cells and produce hydrogen using only water and electricity. The main cost involved with the production of hydrogen in this manner is the price of electricity, and projections suggest the cost of electrolysis excluding electricity will fall to 57.6 p/kg by 2025 [64]. Also, by 2025 the energy required to produce 1 kg of hydrogen will have dropped from 51 kWh to 44.7 kWh [64].

To produce a delivered power of 9270 MWh, assuming an efficiency of 60%, 464000 kg of hydrogen would be required (equivalent to 15450 MWh). Therefore, this equates to a fixed cost of £267,200 plus the electricity cost, as shown in Table 4.

Table 4: Projected cost of 15450 MWh input energy for hydrogen from electrolysis by 2025 (inclusive of the £267,200 fixed cost)

Electricity price (p/kWh)	Projected 2025 price for 15450 MWh of hydrogen (£ Millions)
12.5	2.860
5.0	1.304
2.5	0.786

It is observable that even at the current UK retail price of electricity (around 12.5 p/kWh [65]) then the total cost of the fuel is notably lower than the equivalent value in Table 2. Given the significant decline in cost of solar photovoltaic systems in recent years [66] and the fact that hydrogen could be produced at time of low electricity demand, it is likely that hydrogen producers from 2025 onwards may be able to purchase electricity at a considerably lower price point than currently available. Although future electricity prices are difficult to predict accurately, Table 4 shows that a price point of 2.5 to 5 p/kWh could result in hydrogen fuel being relatively competitively priced. However, it is noteworthy that should the price of hydrogen fall then, as ammonia requires hydrogen to be made, it is probable that the cost of ammonia would also drop.

6.7 WEIGHT OF AMMONIA

Table 2 shows that ammonia has several desirable characteristics. For example, its volumetric density is similar to LNG, and theoretically increasing the fuel oil tank to 119% with ammonia could power the tanker for any journey. Generally, ammonia has shown to have similar characteristics to methanol (CH_3OH) but has the advantage of being carbon free. Also ammonia's price point, despite being 3 to 4 times higher than the currently used fuels, is significantly more competitive than that the other potentially carbon free options (i.e. hydrogen and batteries).

One concern, however, may be the weight of ammonia. Table 2 shows that powering a tanker from only ammonia would increase the total mass of a vessel by over 2.74% compared to LNG when refuelling. An increase in the weight of a vessel has a detrimental effect on performance, therefore more energy may be required to complete the same distances. Conversely, hydrogen is shown to have significantly higher gravitational energy density than any of the power options considered, and may consequently require less fuel than anticipated. It may be valuable to research in more detail the extent that these fuel weights could affect energy demand, such that the figures in Table 2 could be adjusted to take this into account.

In addition, results do not currently account for the additional energy demand of post combustion devices. The use of ammonia would likely require this to reduce NOx emissions.

7. CONCLUSIONS

It is probable that, due to environmental pressures, the shipping industry will be required to move away from oil and gas based fuels. This study has discussed the feasibility of both hydrogen and ammonia for the application of long haul shipping.

Several sources site the low volumetric energy density of hydrogen as a major barrier to it becoming a mainstream fuel source. However, this study suggests that volume requirements of either pressurised gas or liquid hydrogen are not sufficiently high to be considered infeasible. One potential hurdle for hydrogen could be the high price of fuel. However, due to the development of electrolysis technology and the decreasing cost of renewable power, then is possible that hydrogen could be available commercially at a competitive price by 2025.

Ammonia has several advantages over hydrogen, for example less space is required for the same energy content and current market prices are more competitive. However, this study has highlighted one potential concern: ammonia's relatively low gravitational energy density. The fact that this excess weight would likely negatively affect performance should be accounted for in future analysis.

In addition, this study discussed other power sources. It was concluded that batteries are particularly unlikely to meet the size, weight or price requirements for long distance shipping. Also, Methanol appeared to share similar characteristics with ammonia but, as the combustion still produces carbon emissions, it is considered a less desirable option.

In summary, both hydrogen and ammonia have some promising characteristics but there are some significant engineering challenges that need to be addressed before either can be definitively considered either viable or nonviable.

7.1 FURTHER WORK

Several areas have been identified as potential future areas of interest. For example, this paper considered only the volumes of fuels, and not the required storage infrastructure. A detailed investigation into the storage requirements and tank design, particularly for hydrogen, would be valuable. Furthermore, this could be expanded to consider the refuelling process and port-based storage.

The analysis in this study could be further supported by considering data from a range of vessels, particularly for commodity cargo ships as this would be a more probable application for the fuels discussed than a LNG tanker in the short term.

The effect on weight compared to performance could be analysed and integrated into the results of this study. This would be particularly noteworthy given the significant difference between ammonia and hydrogen in terms of gravitational energy density. Figures could also be adjusted to account for auxiliary services and post combustion devices.

8. ACKNOWLEDGEMENTS

The authors gratefully acknowledge financial support from the UK Engineering and Physical Sciences Research Council (EPSRC) through the Centre for Doctoral Training in Energy Storage and its Applications (Grant No. EP/L016818/1).

9. REFERENCES

[1] T. Smith *et al.*, "Third IMO GHG study," 2015.

[2] Department for Transport (2019). *Clean Maritime Plan.*

[3] P. Balcombe *et al.*, "How to decarbonise international shipping: Options for fuels, technologies and policies," *Energy conversion and management,* vol. 182, pp. 72-88, 2019.

[4] L. Jiang, J. Kronbak, and L. P. Christensen, "The costs and benefits of sulphur reduction measures: Sulphur scrubbers versus marine gas oil," *Transportation Research Part D: Transport and Environment,* vol. 28, pp. 19-27, 2014.

[5] H. E. Lindstad, C. F. Rehn, and G. S. Eskeland, "Sulphur abatement globally in maritime shipping," *Transportation Research Part D: Transport and Environment,* vol. 57, pp. 303-313, 2017.

[6] V. Goodright, "Estimates of heat use in the United Kingdom in 2013," December 2014 2014.

[7] N. Drahos. (2019). *LNG Spot Price Forecasting and the Futures Curve.* Available:

https://globallnghub.com/articles/lng-spot-price-forecasting-and-the-futures-curve

[8] M. M. Foss, F. Delano, G. Gulen, and R. Makaryan, "LNG safety and security," *Center for Energy Economics (CEE),* 2003.

[9] K. Kołwzan and M. Narewski, "Alternative fuels for marine applications," *Latvian Journal of chemistry,* vol. 51, no. 4, pp. 398-406, 2012.

[10] K. G. Gkonis and H. N. Psaraftis, "The LNG market: a game theoretic approach to competition in LNG shipping," *Maritime Economics & Logistics,* vol. 11, no. 2, pp. 227-246, 2009.

[11] J. R. Gómez, M. R. Gómez, J. L. Bernal, and A. B. Insua, "Analysis and efficiency enhancement of a boil-off gas reliquefaction system with cascade cycle on board LNG carriers," *Energy conversion and management,* vol. 94, pp. 261-274, 2015.

[12] Royal Dutch Shell plc, "Need for cleaner air driving strong lng demand growth," ed, 2019.

[13] M. A. Korri, "Comparative study of the prospects of hydrogen as an energy source," 2018.

[14] I. Jain, "Hydrogen the fuel for 21st century," *International journal of hydrogen energy,* vol. 34, no. 17, pp. 7368-7378, 2009.

[15] Y. Wang, K. S. Chen, J. Mishler, S. C. Cho, and X. C. Adroher, "A review of polymer electrolyte membrane fuel cells: Technology, applications, and needs on fundamental research," *Applied energy,* vol. 88, no. 4, pp. 981-1007, 2011.

[16] Z. Z. Emission, "One Hundred Passengers and Zero Emissions: The First Ever Passenger Vessel to Sail Propelled by Fuel Cells," ed, 2013.

[17] L. Schlapbach and A. Züttel, "Hydrogen-storage materials for mobile applications," in *Materials for sustainable energy: a collection of peer-reviewed research and review articles from nature publishing group*: World Scientific, 2011, pp. 265-270.

[18] A. P. Pavlos Nikolaidis, "A comparative overview of hydrogen production processes," *Renewable and Sustainable Energy Reviews,* vol. 67, pp. 597-611, 2017.

[19] B. E. Logan, "Ending Our Hydrogen and Ammonia Addiction to Fossil Fuels," ed: ACS Publications, 2019.

[20] A. Bassam, A. Phillips, S. Turnock, and P. A. Wilson, "Design, modelling and simulation of a hybrid fuel cell propulsion system for a domestic ferry," 2016.

[21] J. T. Laine and A. P. Vepsäläinen, "Economies of speed in sea transportation," *International Journal of Physical Distribution & Logistics Management,* vol. 24, no. 8, pp. 33-41, 1994.

[22] M. Hirscher, "Handbook of hydrogen storage," *Topics in applied physics,* vol. 12, 2010.

[23] J. B. Heywood, *Internal Combustion Engine Fundamentals* (McGraw-Hill Series in Mechanical Engineering). New York: McGraw-Hill Book Co, 1988.

[24] C. White, R. Steeper, and A. Lutz, "The hydrogen-fueled internal combustion engine: a technical review," *International journal of hydrogen energy,* vol. 31, no. 10, pp. 1292-1305, 2006.

[25] C. Zamfirescu and I. Dincer, "Using ammonia as a sustainable fuel," *Journal of Power Sources,* vol. 185, no. 1, pp. 459-465, 2008.

[26] J. J. de-Troya, C. Alvarez, C. Fernández-Garrido, and L. Carral, "Analysing the possibilities of using fuel cells in ships," *International Journal of Hydrogen Energy,* vol. 41, no. 4, pp. 2853-2866, 2016.

[27] K. Mazloomi and C. Gomes, "Hydrogen as an energy carrier: prospects and challenges," *Renewable and Sustainable Energy Reviews,* vol. 16, no. 5, pp. 3024-3033, 2012.

[28] A. Züttel, "Materials for hydrogen storage," *Materials today,* vol. 6, no. 9, pp. 24-33, 2003.

[29] S. Makridis, "Hydrogen storage and compression," *arXiv preprint arXiv:1702.06015,* 2017.

[30] C. Sopena, P. Dieguez, D. Sainz, J. Urroz, E. Guelbenzu, and L. Gandía, "Conversion of a commercial spark ignition engine to run on hydrogen: performance comparison using hydrogen and gasoline," *international journal of hydrogen energy,* vol. 35, no. 3, pp. 1420-1429, 2010.

[31] R. F. Service. (2018) Liquid Sunshine. *Science.* Available: https://science.sciencemag.org/content/sci/361/6398/120.full.pdf

[32] N. A. Tim Scarbrough, Aoife O'Leary, Marie Cabbia Hubatova, "Ricardo - Renewable energy's role in decarbonising international shipping," ed: Ricardo, 2019.

[33] G. McLean, T. Niet, S. Prince-Richard, and N. Djilali, "An assessment of alkaline fuel cell technology," *International Journal of Hydrogen Energy,* vol. 27, no. 5, pp. 507-526, 2002.

[34] A. J. Harding, *Ammonia Manufaacture and Uses*. London: Oxford University Press under auspices of Imperial Chemical Industries Ltd. , 1959.

[35] P. J. Crutzen, "The influence of nitrogen oxides on the atmospheric ozone content," *Quarterly Journal of the Royal Meteorological Society,* vol. 96, no. 408, pp. 320-325, 1970.

[36] N. A. Hussein, A. Valera-Medina, and A. S. Alsaegh, "Ammonia-hydrogen combustion in a swirl burner with reduction of NOx emissions," *Energy Procedia,* vol. 158, pp. 2305-2310, 2019.

[37] A. Klerke, C. H. Christensen, J. K. Nørskov, and T. Vegge, "Ammonia for hydrogen storage:

challenges and opportunities," *Journal of Materials Chemistry,* vol. 18, no. 20, pp. 2304-2310, 2008.

[38] D. J. Little, M. R. Smith III, and T. W. Hamann, "Electrolysis of liquid ammonia for hydrogen generation," *Energy & Environmental Science,* vol. 8, no. 9, pp. 2775-2781, 2015.

[39] R. Lan, J. T. Irvine, and S. Tao, "Ammonia and related chemicals as potential indirect hydrogen storage materials," *International Journal of Hydrogen Energy,* vol. 37, no. 2, pp. 1482-1494, 2012.

[40] M. Melvin and T. Amar, "A Review of Ammonia Fuel Cells," *CHE 598 Topic: Fuel Cells & Biofuel cells,* 2015.

[41] M. Intelligence. (2019). *Ammonia market - growth, trends, and forecast (2019 - 2024).* Available: https://www.mordorintelligence.com/industry-reports/ammonia-market

[42] P. Čudina, "Design procedure and mathematical models in the concept design of tankers and bulk carriers," *Brodogradnja: Teorija i praksa brodogradnje i pomorske tehnike,* vol. 59, no. 4, pp. 323-339, 2008.

[43] A. Goldmann *et al.*, "A Study on Electrofuels in Aviation," *Energies,* vol. 11, no. 2, p. 392, 2018.

[44] Elgas LNG. (2015). *Energy Content of LNG.* Available: http://lng.elgas.com.au/energy-content-lng

[45] T. Gresh, *Compressor performance: aerodynamics for the user.* Butterworth-Heinemann, 2018.

[46] W. W. Zittel, Reinhold; Bolkow, Ludwig., "Advantages and Disadvantages of Hydrogen," *Hydrogen in the Energy Sector,* 1996.

[47] Ship&Bunker. (2019). *Rotterdam Bunker Prices*. Available: https://shipandbunker.com/prices/emea/nwe/nl-rtm-rotterdam

[48] U.S. Department of Energy (2015). *Fuel Cells Fact Sheet.*

[49] M. Taljegard, S. Brynolf, M. Grahn, K. Andersson, and H. Johnson, "Cost-effective choices of marine fuels in a carbon-constrained world: results from a global energy model," *Environmental science & technology,* vol. 48, no. 21, pp. 12986-12993, 2014.

[50] Idealhy. (2019). *Liquid Hydrogen Outline.* Available: https://www.idealhy.eu/index.php?page=lh2_outline

[51] J. McKinney, E. Bond, M. Crowell, and E. Odufuwa, "Joint agency staff report on assembly bill 8: assessment of time and cost needed to attain 100 hydrogen refueling stations in California," *Publication Number CEC-600-2015-016,* 2015.

[52] A. Valera-Medina, H. Xiao, M. Owen-Jones, W. David, and P. Bowen, "Ammonia for power," *Progress in Energy and Combustion Science,* vol. 69, pp. 63-102, 2018.

[53] U. Bossel and B. Eliasson, "Energy and the hydrogen economy," *Methanol Institute, Arlington, VA,* 2003.

[54] S. Wittrig. *Ammonia Fuel Opportunities, Markets, Issues*. Available: https://arpa-e.energy.gov/sites/default/files/Wittrig_Ammonia_TransportationFuels_Workshop.pdf

[55] A. Afif, N. Radenahmad, Q. Cheok, S. Shams, J. H. Kim, and A. K. Azad, "Ammonia-fed fuel cells: a comprehensive review," *Renewable and Sustainable Energy Reviews,* vol. 60, pp. 822-835, 2016.

[56] Methanex. (2019). *Methanex posts regional contract methanol prices for North America, Europe and Asia. - Current posted prices*. Available: https://www.methanex.com/our-business/pricing

[57] F. Cerdas *et al.*, "Exploring the effect of increased energy density on the environmental impacts of traction batteries: A comparison of energy optimized lithium-ion and lithium-sulfur batteries for mobility applications," *Energies,* vol. 11, no. 1, p. 150, 2018.

[58] D. L. Wood III, J. Li, and C. Daniel, "Prospects for reducing the processing cost of lithium ion batteries," *Journal of Power Sources,* vol. 275, pp. 234-242, 2015.

[59] M. Steilen and L. Jörissen, "Hydrogen Conversion into Electricity and Thermal Energy by Fuel Cells: Use of H2-Systems and Batteries," in *Electrochemical Energy Storage for Renewable Sources and Grid Balancing*: Elsevier, 2015, pp. 143-158.

[60] L. Ahmadi, S. B. Young, M. Fowler, R. A. Fraser, and M. A. Achachlouei, "A cascaded life cycle: reuse of electric vehicle lithium-ion battery packs in energy storage systems," *The International Journal of Life Cycle Assessment,* vol. 22, no. 1, pp. 111-124, 2017.

[61] G. Berckmans, M. Messagie, J. Smekens, N. Omar, L. Vanhaverbeke, and J. Van Mierlo, "Cost projection of state of the art lithium-ion batteries for electric vehicles up to 2030," *Energies,* vol. 10, no. 9, p. 1314, 2017.

[62] S. Rajagopalan, A. Maitra, J. Halliwell, M. Davis, and M. Duvall, "Fast charging: An in-depth look at market penetration, charging characteristics, and advanced technologies," in *2013 World Electric Vehicle Symposium and Exhibition (EVS27)*, 2013, pp. 1-11: IEEE.

[63] J. D. Holladay, J. Hu, D. L. King, and Y. Wang, "An overview of hydrogen production technologies," *Catalysis today,* vol. 139, no. 4, pp. 244-260, 2009.

[64] E. Shafiei, B. Davidsdottir, J. Leaver, H. Stefansson, and E. I. Asgeirsson, "Energy, economic, and mitigation cost implications of transition toward a carbon-neutral transport

sector: A simulation-based comparison between hydrogen and electricity," *Journal of cleaner production,* vol. 141, pp. 237-247, 2017.

[65] U. Power. (2019). *Gas & Electricity Tariff Prices per kWh.* Available: https://www.ukpower.co.uk/home_energy/tariffs-per-unit-kwh

[66] S. Reichelstein and M. Yorston, "The prospects for cost competitive solar PV power," *Energy Policy,* vol. 55, pp. 117-127, 2013.

9. AUTHORS BIOGRAPHY

Charles J McKinlay holds the current position of PhD researcher at the University Of Southampton, as part of the CDT in Energy Storage and its Applications programme. His research is focused on Future Fuels for Shipping. His background includes Mathematics and Energy Engineering.

Professor Stephen Turnock, FRINA is Head of Civil, Maritime and Environmental Engineering at the University of Southampton. He has long standing interests in decarbonisation of shipping and in particular analysis and modelling of alternative power trains including the use of hybrid battery systems.

Professor Dominic Hudson is the Shell Professor of Ship Safety and Efficiency at the University of Southampton. He has research interests in all aspects of ship hydrodynamics, particularly ship design and operation for energy efficiency. He has published over 140 papers in journals and proceedings of refereed conferences. He is a Chartered Engineer and member of RINA. He is presently a member of the 29th ITTC Specialist Committee on Ships in Operation at Sea and the RINA IMO Committee.

A REVIEW OF MOORING SYSTEM SOLUTIONS FOR FSRUs.

G. Mortola and **J. Dunstan**, London Marine Consultants Ltd., UK

SUMMARY

The energy industry is moving towards a low carbon future. During this paradigm shift, natural gas has taken its place as a transition fuel with a forecasted demand growth for the next 20-40 years. Liquefied Natural Gas (LNG) is the preferable transportation method for natural gas over long distances, allowing producers to reach new markets and nations to achieve energy independence. FSRUs (Floating Storage Regasification Units) are key components in the LNG supply chain and are essential for the economics of projects that require low CAPEX and short delivery schedule that cannot be achieved with similar onshore facilities.

One of the critical components of an FSRU is the mooring system. FSRUs can be located on jetty side, nearshore or even far away from shore in typical offshore environments, therefore there is no standard choice of mooring system and designers must select the mooring system type on a project specific basis.

This paper provides a review of the different mooring solutions for the current and future FSRU fleet, introduces the available options and for each it outlines the typical benefits and shortcomings. As a result of the critical review of the mooring system options a schematic decision-tree is provided, which allows FSRU owners, operators and port authorities to identify the optimal mooring solutions based on project specific constrains and characteristics. The paper also discusses some case studies focusing on the mooring system selection, design, construction, integration and installation for FSRU projects.

NOMENCLATURE

ABS	American Bureau of Shipping
API	American Petroleum Institute
CAPEX	Capital Expenditure
FSRU	Floating Storage Regasification Unit
HMPE	High Modulus Polyethylene
LNG	Liquefied Natural Gas
NM	Nautical Mile
OCIMF	Oil Companies International Maritime Forum
O&M	Operation and Maintenance
YRP	Year Return Period

1. INTRODUCTION

1.1 NATURAL GAS MARKET OUTLOOK

In the past 10 years, the rise of renewable energy has occupied the spotlight of energy related news, demonstrating the shift towards a "greener" low carbon future. This change is expected to be a slow transition more than a revolution, lasting more than 30 years [9]. At the same time, the overall energy demand from fossil fuels (coal, gas and oil) will still rise albeit at a slower pace, with only coal showing a visible sign of decline.

Figure 1 from [1] shows the outlook, up to 2040, of primary energy consumption by fuel (Figure 1-a) and share of primary energy fuel type (Figure 1-b). The figure shows that renewables and natural gas are the only energy sources that will increase their respective share and importance in the next 30 years.

Natural gas will take the place of "transition fuel" filling the gap created by the overall small growth in demand of coal and oil as energy sources combined with the growth in renewable sources. This change will require the development of a new supply chain that permits a rapid and economical distribution of natural gas on a global scale.

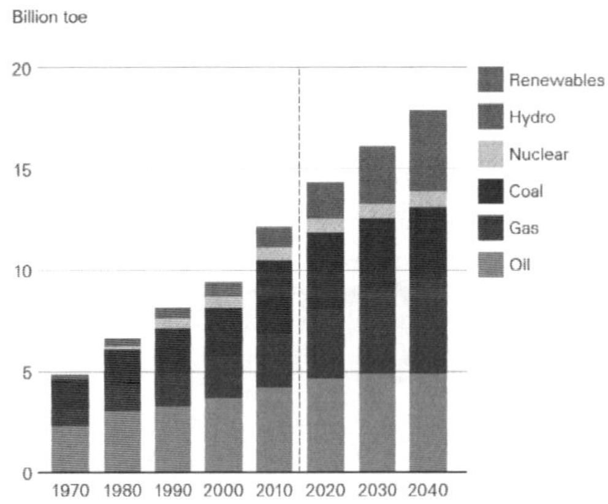

(a)

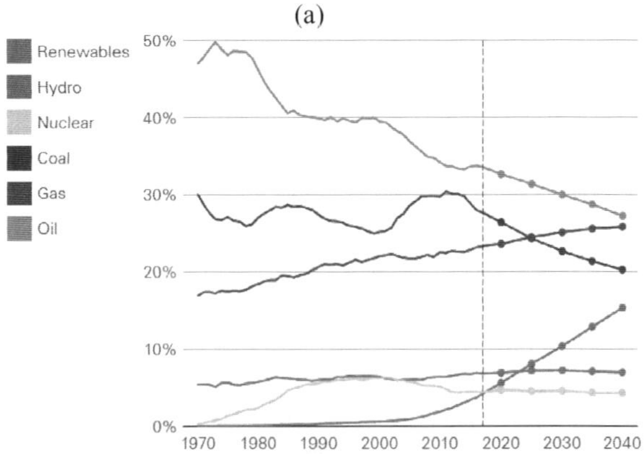

(b)

Figure 1: (a) Primary Energy Consumption by Fuel, (b) Share of Primary Energy

1.2 FSRUs AND THEIR POSITION IN THE NATURAL GAS SUPPLY CHAIN

The transportation of natural gas generally occurs via pipelines, located either on land or offshore, or via ship transport in liquefied state, referred to LNG (Liquefied Natural Gas). The selection of the method of transport depends on many factors such as distance from source to customers, duration and volume of the supply, available infrastructures and political/national interests.

[2] compares the cost of gas transport between pipeline and LNG for different scenarios such as onshore/offshore pipelines, volume of delivery, etc. Although the conclusions of the comparison are difficult to generalise, they indicate that LNG transport via ship appears to be a more economical means of transport for natural gas over distances longer than 3,500NM compared to onshore pipelines and 700NM for offshore pipelines.

In general, transport of natural gas as LNG requires the natural gas to be liquefied by cooling down (to -163°C) at the port of departure, transported in purpose made vessels (i.e. LNG carriers), regasified at the port of arrival and distributed to the end users via pipeline. FSRUs enter the supply chain as port of arrival and regasification facilities.

FSRUs are designed to receive and store LNG from trading LNG carriers, converting it into gas state and delivering it to customers. Usually FSRU design is based on conventional LNG carriers with the inclusion of regasification and metering modules. This design allows FSRUs to be also deployed as trading LNG carriers. Some FSRUs of smaller capacity, which cannot provide an economic opportunity to be utilised as both trading LNG carrier and FSRU are built based on barge design without a propulsive system.

Figure 2 (a) shows a typical FSRUs based on LNG carrier with prismatic tanks, while Figure 2 (b) shows a smaller barge FSRU design (b).

Usually smaller FSRUs as the one depicted in figure 2 (b) are required for projects in which the natural gas is needed as fuel for a single consumer, such as an electric power station or to supply isolated communities.

(a)

(b)

Figure 2 (a) Typical FSRU Based on LNG Carrier Design, Courtesy of MOL, (b) Barge Shape FSRU, Courtesy of PaxOcean

A FSRU, compared to an onshore based terminal, requires a smaller physical footprint; the natural gas is stored and regasified on board requiring a significantly smaller onshore infrastructure.

The necessity of building less onshore infrastructure also translates in a smaller project schedule, which can further be reduced if a FSRU is idle available in the market.

Another advantage of FSRUs is the reduced CAPEX compared to a conventional onshore terminal. Usually the contracting strategy for these units is on a lease/operate basis, where a shipping company leases the vessel at an agreed daily rate significantly reducing the necessary upfront cost at project start-up.

In order to be economically viable a FSRU needs to be located as close as possible to the gas distribution network or the final gas end user, for example a factory or power plant. This is done to reduce the amount of additional infrastructure needed to deliver gas: for example, if a FSRU is moored away from the shore there is the necessity to install a subsea pipeline which requires significant additional cost that can compromise project economics. For developed countries, this usually translates to the FSRUs being located in a sheltered commercial port. For developing countries or remote locations, where port infrastructures may not be present, FSRUs could be moored in more exposed areas away from typical shipping routes and navigation channels. These locations can either be near shore on a purpose built jetty or offshore.

1.3 INTRODUCTION TO MOORING SYSTEMS FOR FSRUs

The type and composition of the mooring system of a FSRU varies significantly on a project specific basis.

The majority of FSRUs are moored to a jetty-side, in a mooring configuration typical of a ship visiting a harbour quayside. On the other hand, some vessels are moored via submerged disconnectable buoy, weathervaning turret or yoke mooring systems.

Some FSRUs are designed to be permanently moored and the mooring system must safely maintain the vessel in station during the entire chartered period. The results are stringent design conditions, usually based on 100- Year Return Period (YRP) events. Other mooring systems are defined as temporary moorings and designed to keep the FSRU on station up to a specific set of design environmental conditions, defined as "FSRU connected" condition. If the weather is forecasted to exceed the "FSRU connected" condition the vessel starts the disconnection procedure, disconnects from the mooring system and sails away from the storm in either open seas, or to a sheltered location.

The variety of mooring system configurations available highlights the importance of the selection of the appropriate stationkeeping system.

The main difference between the mooring system of a FSRU and offshore oil and gas unit is the fact that a FSRU is frequently moored in shallow waters as close as possible to the shore.

Catenary mooring systems are the standard mooring system configuration for offshore floating facilities in shallow and medium water depths. In very shallow water, (i.e. less than 20m water depth) typical catenary mooring systems are not very effective, this is because the moorings are unable to obtain the restoring force from the mass of the line lifted from the seabed. Therefore, in very shallow water, to maintain the vessel within an offset that does not compromise the integrity of the flexible pipes which export gas to the shore, a significant mooring hardware mass is needed to be lifted, resulting in large chain diameters and elevated mooring line numbers.

The nearshore position of a FSRU, the sloping seabed bathymetry, geographical uniqueness of each location and passing ship traffic are examples of some phenomena specific to FSRU design which create an additional set of loads that must be considered during the design of the mooring system.

The above described characteristics and effects represent a unique set of challenges that need to be addressed during the design of the mooring system. The resulting technical solutions can therefore diverge from typical offshore and marine applications.

This paper explores all the possible mooring solutions available to FSRU designers; the different mooring options are presented, advantages and shortcomings discussed, and design considerations and criteria highlighted. It also conducts a review of the mooring system configurations utilised by the current floating gas terminals. Finally, the overall aim of this paper is to provide the reader with a set of information and tools that will allow to critically judge and select the optimal mooring system configuration for a FSRU.

2. REVIEW OF MOORING SYSTEM OPTIONS

2.1 GENERAL

This Section provides a description of the different mooring system options available to FSRU designers. The review is conducted for the following mooring system types:

- Jetty side moorings;
- jetty islands;
- turret mooring systems (permanently moored);
- turret mooring systems (disconnectable);
- yoke systems;
- spread mooring systems.

Other solutions for regasification terminals such as gravity base structures (LNG Rovigo) or jetty-less LNG transfer systems (LNG Connect) are outside the scope of this review.

2.2 JETTY SIDE MOORING

Jetty mooring is the typical mooring option for a quayside moored vessel, Figure 3 shows a typical jetty mooring arrangement. In a jetty mooring arrangement, the vessel is moored alongside a jetty structure and the mooring lines run from the fairleads on the vessel main deck to anchor points on the quay or jetty.

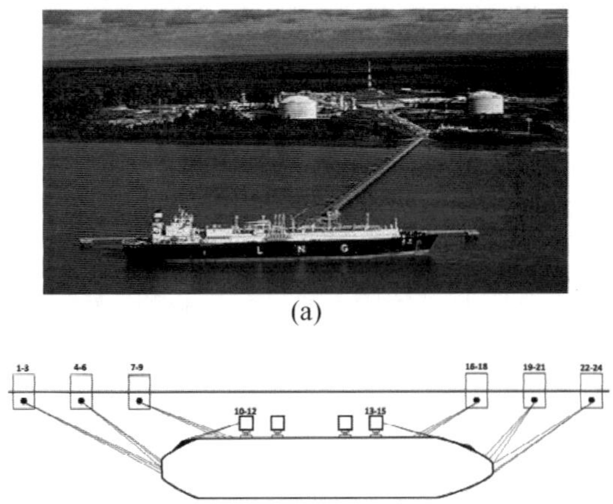

(a)

(b)

Figure 3. Typical Jetty Mooring Configuration, (a) Layout, (b) Mooring Line Schematic [6]

The typical mooring layout can be seen in Figure 3(b). Mooring lines are grouped into two categories based on the motion they need to restrain: breast lines and spring lines. Breast lines run from the bow and stern of the vessel to the jetty, these lines run perpendicular to the vessel side as their objective is to restrain the lateral movement of the ship. Spring lines run parallel to the vessel side departing from fairleads located close to the midship. The spring lines limit the longitudinal movement of the FSRU. In some cases, the mooring system includes other sets of lines, namely head and stern lines to further limit movement of the vessel.

Mooring lines can be made of man-made fibre such as HMPE, polyester, nylon or even steel wire. The material selection depends on several factors such as type of loading, length of the mooring lines, vessel size and allowable maximum vessel excursion. Lines in the same group (e.g. forward breast lines) should have similar characteristics and pretension, to allow for a more uniform load sharing. Lines made up of less stiff material (e.g. nylon) allow the vessel to move more by reducing peak dynamic loads, but the larger motions displacement could result in damages to the gas export system.

On the jetty side the mooring lines terminate to mooring bollards or quick release type connectors, which allow for load monitoring and rapid disconnection in case the maximum mooring working loads are exceeded.

Although the geometrical layout of a jetty side mooring system for a FSRU is similar to the one of a normal trading vessel, the design requirements are significantly different. For a trading tanker jetty mooring is seen as a temporary activity and its design is usually based on static load balance calculations following industry guidelines, as the OCIMF guidelines in [7]. The mooring system for a jetty side FSRU is considered a long-term system and therefore subjected to more rigorous design conditions based on classification societies rules, such as [6] and [10]. Class requires dynamic mooring analysis and fatigue studies based on a more detailed hydrodynamic modelling of the hull. Effects such as passing vessels, ship to ship and ship to pier interaction and local bathymetry effects must be included. Class rules only apply to the vessel side of the mooring system, the design of mooring hooks and bollards on the pier and the its structure are regulated by state/port regulation and industry guidance, therefore care should be taken to ensure consistent equipment specifications and strength capacity of the different mooring components.

The main advantages of a jetty mooring systems can be defined as follows:

- It is established and well-known technology;
- The construction works can be performed by a large pool of civil contractors, providing also the possibility to satisfy local national content requirements in developing countries;
- The jetty provides easy access to the FSRU for both personnel and supply;
- The marine facilities on the vessel and pier are less exposed to the environment compared to vessel moored in open water, resulting in less wear and Operation & Maintenance (O&M) costs;
- Facilities are connected to shore.

The main disadvantages are:

- If existing structures are not available, the construction of a purpose made jetty requires a significant amount of civil works and an associated long delivery schedule;
- To make the area accessible to large LNG carriers and to the FSRU itself dredging maybe required both during the construction of the terminal and during its operation, to clear the seabed of any possible sediments;
- Jetties located in navigation channel subjected to large traffic provide a navigation obstruction and possible hazard;
- The proximity to the shore can create concern to local communities and requires a longer approval process;
- All pipework connecting the FSRU to natural gas tanks or network are usually exposed to the open increasing the security risks.

2.3 JETTY ISLAND

Jetty island moorings are jetty like structures that are located away from the shore, with no surface connection to it. The FSRU is moored to the jetty island in the same manner as to a shore based mooring jetty. Figure 4 shows a typical layout of a jetty island.

Figure 4 Jetty Island Mooring, Courtesy of Excelerate Energy

Considerations on mooring design criteria and requirements are similar to the ones outlined in the previous section for jetty side moorings. The main difference between the two mooring options is the fact that for the jetty island some level of subsea infrastructure will need to be installed to transfer natural gas to the shore, requiring a specialised contractor and different contracting strategy compared to shore based jetty.

The main advantages for jetty island compared to shore based jetty side mooring are the followings:

- Jetty islands permit to move the terminal further away from the coast in more accessible water, reducing the amount of dredging required;
- Jetty islands present a reduced impact on marine traffic since they can be located outside shipping routes;
- Smaller footprint compared to shore jetties. For this type of system personnel can only access by boat, therefore the amount of structure can be reduced since there is no need to provide road access. For example, in some systems the mooring bollards are located on separate

structures not interconnected, significantly reducing the amount of foundation piles and walkways needed (jetty pile or dolphin pile system);

- Increased security due to the limited accessibility and exposure of key elements (e.g. pipeline);
- Smaller visual impact for local communities.

On the other hand, jetty islands present the following shortcomings compared to shore-based terminals:

- Larger O&M costs due to the remote location and need to boat transfer for personnel and material;
- Jetty islands require additional cost due to the subsea pipeline needed to connect the terminal to the shore.

2.4 WEATHERVANING MOORING SYSTEMS

In case the FSRU needs to be moored further offshore, sea island jetty moorings are not generally a feasible solution. The FSRU will be located in deeper waters and it will eventually be exposed to harsher environmental conditions. The construction of pile jetty in water depths larger than typical harbour depths could be economically unfeasible, and the utilisation of a jetty like mooring arrangement in harsher environments could result in a large number of mooring lines and a reduced terminal availability due to the possibility that the FSRU has to disconnect from the moorings and sail away in case of a storm. In this case, a weathervaning mooring system solution is the preferable solution.

In a weathervaning mooring system the vessel is free to rotate around a point, defined as the turret, to align with the prevailing weather and reduce environmental loads and vessel motions/responses. The weathervaning mooring systems available for FSRU projects can be grouped into two categories:

- Turret mooring systems, and
- Yoke mooring systems.

In a turret mooring system the turret is attached to the vessel and connected to the seabed via mooring lines. In a yoke mooring system, the centre of rotation is on an external structure rigidly connected to the seabed (e.g. via jacket tower) and the vessel is connected to it via a counterweight system called yoke. Figures 5(a) and (b) show respectively a picture of a bow column turret and a image of a tower yoke system for FSRUs.

Some turret mooring systems are disconnectable. In this case, the FSRU can disconnect from the mooring to either reload LNG or sail away in case of a storm.

The design of these mooring system types is well known and proven in the industry. The station keeping system of the vessel has to comply with the same classification societies rules for offshore floating production facilities. See [6] as example. Offloading of LNG to a turret or yoke moored FSRU still occurs in a side by side configuration as per jetty side moorings. The weathervaning capacity of the FSRUs can be used to shelter the side from which LNG transfer will occur resulting in an increase of terminal availability

2.4(a) TURRET MOORING SYSTEM

The mooring lines for a turret system in shallow water depths (less than 80-100m) are in general catenary type made up of studless mooring chain.

(a)

(b)

Figure 5 (a) External Column Turret, FSRU Toscana, (b) Tower Yoke System, Lampung FSRU

The bottom structure of the turret is called chain table, this is where the mooring lines terminate and the flexible risers, that carry hydrocarbons and service fluids, connect to the structure. The turret is connected to the vessel via a single or a series of bearings that uncouple the turret to the yaw rotation of the vessel. The transfer of products between the vessel and the turret occurs via a swivel system located at the top of the turret. Figure 6 shows an example of toroidal swivel for turret mooring applications.

Figure 6 Example of Toroidal Swivel for Offshore Oil & Gas Industry

Turret mooring systems are defined based on the position of the turret within the vessel. Systems with turret located in a moonpool inside a vessel's hold are called "Internal

Turret System". Mooring systems with turrets on the exterior of the hull, typically at the bow, are named "External Turret System". Figure 7 shows a schematic comparison between internal and external turret systems.

External turret systems can be further classed based on the position of the turret chaintable. If the chaintable is at keel level the turret is called "External Column Turret System", if the chaintable is at main deck elevation and is connected to the hull using a cantilever like structure they are called "External Cantilever Turret System". Figure 8 shows a comparison between external column and cantilever turrets.

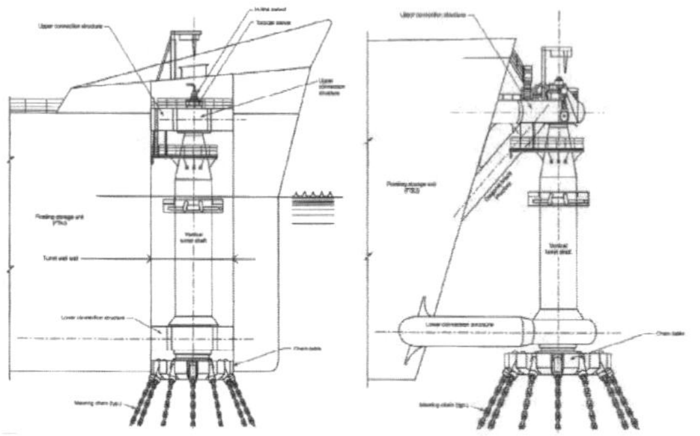

Figure 7 Internal (left) and External (right) Mooring Turret Systems, from [8]

Figure 8 Example of External Column Turret (left) and External Cantilever Turret (right)

For FSRU applications a column turret is often preferred to the external cantilever turret option, this is because FSRUs are based on LNG carrier design and the fore end shape of this vessel type is better suited to be integrated with a column like structure that connects at the keel and main deck level (see Figure 8). External cantilever turrets that connect to the hull at main deck elevation require a blunt bow for a better load sharing / structure integration.

Installing an external turret system on a FSRU entails major alteration work and requires modifications that will prevent the vessel to be easily utilised as a trading LNG carrier. Therefore, these mooring solutions are better suited for long chartered durations, typically more than 5 years, new purpose-built units or for vessels at their last deployment.

The main advantages of turret mooring systems compared to previous described systems are the followings:

- They are a well-known established technology;
- They allow the FSRUs to be moored offshore;
- No upper bound limit on water depth;
- The weathervaning allows the vessel to be moored in area with no relevant environmental directions;
- If the FSRUs is equipped with a stern thruster, heading can be adjusted to improve operability for offloading operations;
- The vessel can be moored away from urban centre simplifying the review process.

The main shortcomings of turret mooring systems are as below:

- Turret mooring require swivel systems that may increase CAPEX;
- Both internal and external turret systems require vessel modifications and drydocking;
- Lower bound limit in water depth (30-40 meter).
- Moving the FSRU away from the shore results in a CAPEX increment for pipeline construction.

2.4(b) DISCONNECTABLE TURRET MOORING SYSTEM

Most turret mooring system utilised by FSRUs are of the disconnectable type (see Section 3). In a disconnectable turret mooring system the mooring lines and flexible risers terminate to a subsea buoy. The FSRU, once it arrives on the mooring location, hooks-up to the buoy and lifts it into a moonpool at the fore end of the vessel. Once inside the hull, the buoy is locked in position and flowlines connected to the swivel, after this the vessel can freely weathervane around the turret buoy.

Although several types of disconnectable turret buoy system exist in the market, the standard type of system for FSRU application is the internal submerged turret buoy, as shown in Figure 9.

Figure 9 Disconnectable Turret Buoy Mooring System, Moheshkhali Floating LNG

Disconnectable turret mooring buoy systems can only be utilised with FSRUs that have been specifically designed and built with the space and equipment to accommodate a mooring buoy in the fore hold.

Since, in case of severe weather, the FSRU by design disconnects from the moorings, the mooring lines size is smaller compared to permanent mooring turret units. This also results in the possibility to install a disconnectable turret system in shallower water depth, since the excursion that the riser system must accommodate are small compared to a permanently moored turret option.

The following advantages, in addition to the ones presented in Section 3.3(a), can be associated with a disconnectable turret mooring system:

- FSRUs can be moored in areas prone to cyclones or tropical storms with smaller mooring hardware, compared to equivalent permanent option;
- FSRUs can be moored in shallower water depths due to the reduced environmental design conditions;
- One FSRU can connect to buoys with different mooring line numbers, in different water depths and with different export riser configurations with minimal modification works.

The main drawbacks of disconnectable turret mooring systems, compared to permanent turret option are as follows:

- Only FSRUs specifically designed to receive a turret buoy can utilise this system;
- Mechanical equipment for mooring buoy hook-up and fluid transfer results in a more complex and expensive option compared to traditional permeant turret mooring systems;
- Only a small number of suppliers can provide a disconnectable turret system.

2.4(c) YOKE MOORING SYSTEM

In benign environments, yoke systems are a feasible solution for water depths smaller than circa 30 meters. For these shallow water depths, the design of a turret system can be challenging since the small dimension of the water column reduces the effectiveness of the mooring lines to generate sufficient restoring force to maintain the FSRU within an small excursion envelope, as described in Section 1.3. The design of the export riser system is also critical in shallow water: the riser system needs to be compliant and accommodates for the vessel excursion, but also needs to provide positive clearance with the FSRUs keel when the vessel swings above the risers, while weathervaning. Therefore, riser and mooring systems constitute a limit to define a lower bond water depth value below which it is either not feasible or economical to consider a turret mooring system.

In a yoke system the turret is fixed with respect to the ground either above water on a jacket structure (Tower Yoke) or directly on the seabed on top of a foundation (Submerged Yoke). The FSRU is then connected to the turret via an articulated counterweight system (yoke) which incorporate a set of bearings and uni-joints to uncouple the FSRU motions from the turret. Yoke systems are defined as "Rigid", if the yoke is connected to the vessel via rigid elements of "Soft" if the connection is made up of chain or wire. Figure 5(b) shows an example of Rigid Tower Yoke system, while Figure 10 shows Submerged Soft Yoke system.

Figure 10 Submerged Soft Yoke System (left), Courtesy of NOV

The restoring force is provided by a discrete ballast system instead of long chain section, for this reason this mooring system type can be utilised at smaller water depths compared to mooring turret systems.

The main advantaged of a yoke system compared to turret moorings are the listed below:

- Yoke systems can be utilised in shallow water depth not accessible to conventional mooring turret systems;
- The modifications required on the vessel are at main deck level, therefore there is no need for drydocking.

On the other hand, a yoke mooring system presents the following downsides:

- Cost of a yoke mooring system can be significantly larger compared to turret mooring systems;
- Only a small number of suppliers can provide a disconnectable turret system.

2.5 SPREAD MOORING SYSTEMS

For projects in which the FSRU must be located away from the shore in areas characterised by environment with a prevalent directionality a spread mooring system can be a technical and economically feasible solution. Spread mooring systems are the simplest form of long term mooring available to offshore units, mooring lines are connected on the vessel side either at keel or main deck elevation. Figure 11 shows two examples of spread mooring systems.

(a)

(b)

Figure 11 Spread Mooring Systems for Offshore Production Unit (a) and FSRU (b)

In a spread mooring system, the heading of the vessel is restricted towards the prevailing environmental direction, therefore there is no need for a swivel to connect the vessel to the export flowlines, as per turret mooring system.

The modifications required to utilise a spread mooring system are minimal, the vessel side or deck needs to be reinforced in the areas where the chain stoppers will be installed, mechanical equipment for mooring installation needs to be integrated together with piping for transferring fluid from FSRU regassification system to the export riser system.

Although spread mooring systems are a practical solution that could help improve economic feasibility for some projects, they have thus far never been utilised with FSRUs. This is mainly due to the difficulty to perform ship to ship LNG transfer with conventional spread mooring systems, since the mooring lines running from the side towards the seabed do not permit a LNG carrier to approach the side of the FSRU. Some advancement has been proposed in recent years with positioning of the chain stopper at the FSRUs keel and with interconnected catenary mooring lines. At the present time the only project, at design phase that is considering utilising a spread mooring system is the El Salvador FSRU.

The list below provides a summary of the main advantages of spread mooring systems:

- Spread mooring systems require minimum modification cost and time;
- Simple technology that do not require expensive mechanical equipment (e.g. swivel);
- Spread moored vessels can accommodate for a large number of risers.

While the main disadvantages are the following:

- Spread mooring systems with chainstoppers above water present a mooring integrity risk for offloading LNG carriers approaching the side of the FSRU;
- FSRU cannot change heading to sheathed LNG carrier during offloading resulting in a reduced terminal availability.

3. MOORING SYSTEMS FOR FSRUs, PAST AND PRESENT FLEET

A review of the mooring system solution utilised by the past and current FSRUs fleet has been conducted based on [3], [4] and [5]. The review included the past and current FSRUs projects, therefore multiple deployments of single vessels have been considered as different projects.

Figure 12 shows a summary of the different mooring system type utilisation for the different FSRUs projects.

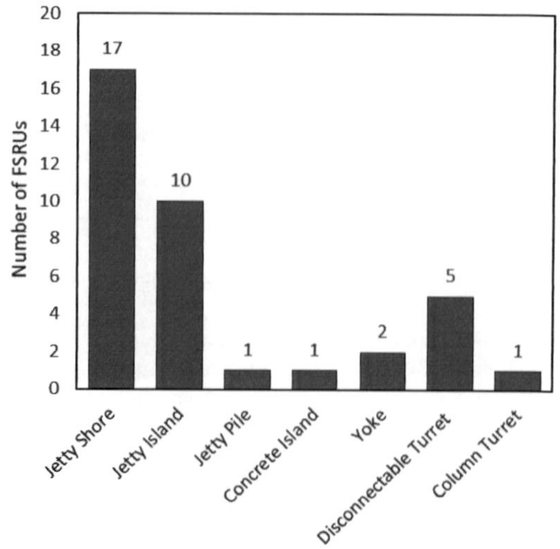

Figure 12 Mooring System Types for Past and Present FSRUs Fleet

The figure shows that the majority of FSRUs have been deployed within a commercial port or in its surroundings and moored to a jetty either connected or not connected to the shore (jetty island and jetty pile). In circa 20% of cases, FSRUs are moored using a weathervaning mooring system. There is only one system, the Adriatic LNG Rovigo, for which the FSRU is actually a gravity base structure.

By looking at the characteristics of the projects that utilise a weathervaning mooring system, either turret or yoke, it can be seen that these types of systems allowed FSRUs to be moored in large water depth range up to 112m (FSRU Toscana) and in environments subjected to hurricanes and cyclones (for disconnectable turret systems).

The review also demonstrated that no FSRU utilises a spread mooring system. The only exception is the

Electrogas FSU, in Malta, in this case a spread mooring system is utilised as "storm mooring". During normal operation, the vessel is jetty moored, but in case of a storm approaches the island the vessel disconnects from the jetty and connects to a chain based spread mooring system in the vicinity, for this reason this system has been classified as jetty moored in Figure 12.

4. FSRU MOORING SYSTEM SELECTION

The consideration from the mooring system option review presented in Section 2 has been utilised to create a flow chart to assist FSRU operators, port authorities or any person not familiar with details of mooring system design in the review and initial selections of a mooring system types for an FSRU project. The flow chart presented in Figure 13 also provides a project name as an example for each mooring option presented. In case there are no current operating or proposed examples, they are listed as "none".

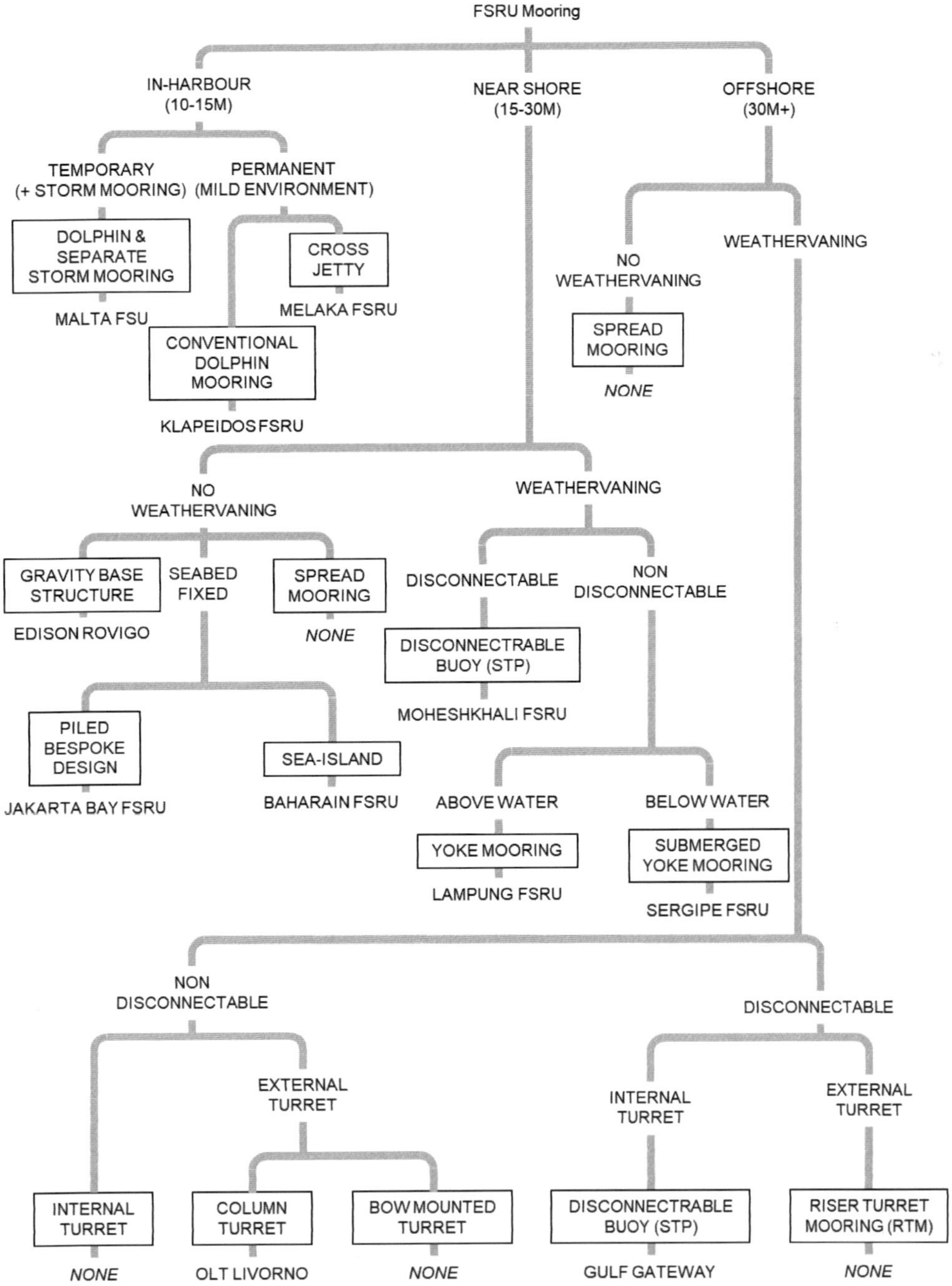

Figure 13 Flow Chart for Moring System Type Selection

5. CONCLUSIONS

This paper presents the mooring system options available for FSRU systems. The range of mooring system types utilised highlights the difference between each FSRU project, from jetty side ship or barge to offshore vessel moored using complex turrets or yoke systems.

The different mooring system types have been presented, discussed and the main advantages and shortcomings listed. The findings of this review have been utilised to create a flowchart that guides the selection of mooring system options for a specific FSRU project.

The paper demonstrates the importance of the correct mooring system selection, especially when conventional jetty side moorings are not available or economically feasible to build. In this cases, the utilisation of mooring system types typical of offshore Oil & Gas industry can open the possibility to develop projects that otherwise could not be technically and commercially feasible.

6. REFERENCES

1. British Petroleum, 'BP Energy Outlook, 2019 Edition', BP plc., 2019
2. S. Cornot-Gandolphe et al., 'The challenges of further Cost Reductions for New Supply Options (Pipeline, LNG, GTL)', 22nd World Gas Conference, 1-5 June 2003.
3. B. Songhurst, 'The outlook for Floating Storage and Regasification Units (FSRUs)', The Oxford Institute for Energy Studies, July 2017.
4. B. Songhurst, 'Floating LNG Update – Liquefaction and Import Terminals', The Oxford Institute for Energy Studies, September 2019.
5. DataFusion Associates, 'FSRU's – the Great Gas Game Changer', DataFusion Associates, February 2017.
6. ABS, 'Guidance Notes on Nearshore Position Mooring', *ABS*, February 2019.
7. OCIMF, 'Mooring Equipment Guidelines (MEG 4)', OCIMF, 2018.
8. American Petroleum Institute, 'Design and Analysis of Stationkeeping Systems for Floating Structures', American Petroleum Institute, October 2005.
9. DNVGL, 'Energy Transition Outlook 2019', https://eto.dnvgl.com/2019.
10. DNVGL, OTG-18 Guidance for Long-term Nearshore Mooring, January 2019

7. AUTHORS BIOGRAPHY

Dr. Giuseppe Mortola holds the current position of Lead Naval Architect at London Marine Consultants Ltd. (LMC). He co-ordinates the design of mooring systems and is responsible for the naval architecture scope of LMC projects. Giuseppe obtained his Ph.D. in 2012 from Universities of Glasgow and Strathclyde on the topic of non-linear wave ship interactions. His experience includes design of mooring systems, hydrodynamic studies, model tests and hull design for different floating structures.

Jon Dunstan holds the current position of Managing Director at LMC. He is responsible for overall operations and business development within the organisation having been involved with the company for 22 years. He holds a BEng in civil engineering from Imperial College London.

LNG BUNKER VESSEL DESIGN AND OPERATION

K. W. Hutchinson and **D. C. Dobson**, Safinah Ltd., UK

SUMMARY

Previous papers by the authors to RINA International Conferences held in 2016 [1], 2017 [2] and 2018 [3], articles in RINA magazines [4 to 6], together with an earlier paper [7], have discussed both large deep-water offshore and small-scale near-shore floating Liquefied Natural Gas (LNG) assets in conjunction with it transoceanic and coastal transhipment between production and reception hubs. This paper builds upon these by turning attention to the emerging LNG bunkering sector. For more than a century our world has revolved around fossil fuels, but the transformation from this via transitory fuels to zero carbon fuels is happening now and faster than anybody could have predicted a few years ago. LNG is arguably the transitory fuel of choice for both short-sea coastal and deep-sea ocean-going shipping. Hence, recent years has seen the introduction of a new class of ship namely the LNG Bunker Vessel (LNGBV or simplified to LBV) or sometimes termed a Gas Supply Vessel (GSV). Drawing on nearly two decades of experience in the design and evaluation of numerous new-build design, conversion and upgrades of offshore production and regassification assets together with a decade of experience in the design of conventional and highly novel and innovative LNG and other Gas Carriers, of all capacities and configurations, the authors consider the development and robust design of LBVs for deployment worldwide.

1. THE HISTORY OF LNG SHIPPING

From the initial concepts from 1950 in the United States of America (USA), the development of LNG shipping has grown significantly to its current day size and scale. The earliest LNG Carrier (LNGC) carried only 2% of the cargo of the largest ship in the fleet of 2020. Today the LNGC fleet numbers 525 ships, with a further 118 under construction.

Facility and ship development were initially very slow and limited to certain trade routes and countries, transatlantic from the USA to the United Kingdom (UK) being the first. However, there are currently 19 exporting countries and 38 importing countries. There were in excess of 5,200 trade voyages in 2019.

1.1 WORLD'S FIRST LNGC

The oceangoing transhipment of LNG can be traced back over six decades to late 1950s, when a former Second World War built cargo ship was converted to transport LNG in bulk from the USA across the Atlantic Ocean to the UK.

The ship in question was the *MS Marline Hitch* which was an 11 knot single-screw Type C1-M-AV1 general cargo ship. The ship was converted to an LNGC in 1958 by the Alabama Drydock and Shipbuilding Company in Mobile, Alabama, and renamed *MS Methane Pioneer*. In order to enable safe transportation of the cryogenic cargo, storage tanks constructed of aluminium were installed with a total storage for over 2,300 tonnes of LNG, a capacity of 5,088 metres^3 (32,000 barrels), and as-converted she had a revised GRT of 4,693 and deadweight capacity of 4,907 tons.

She transported the world's first ocean cargo of LNG in early 1959, from Louisiana USA to Canvey Island UK. She made a total of 30 voyages between 1959 and 1972 when she was converted into a storage vessel for LNG before being broken-up later that year.

1.2 WORLD'S FIRST NEW-BUILD LNGCs

Following on from the first shipping of LNG across the Atlantic, described above, the first two purpose-built LNGCs were built in the UK for British Gas in the early 1960s for shipping of LNG from Algeria.

The ships in question were the 15.5 knot 10,25 MW (13,750 shp) steam turbine powered *SS Methane Princess* and *SS Methane Progress*. The ships had nine Conch independent type aluminium LNG cargo tanks giving a capacity of 27,400 metres^3, on a Length Overall (L_{OA}) of 189 metres (621 feet) and Breadth (B) of 25 metres (82 feet).

Figure 1: 27,400 metres*3* LNGC *SS Methane Progress*, built in 1964

1.3 WORLD LPGC / LNGC FLEET

As of January 2020, the world fleet of Liquefied Petroleum Gas Carriers (LPGCs) currently in operation numbers 1,474, with approximately a further 120 under construction. Of these, the number of which are in excess of 50,000 metres³ capacity, and hence termed as Very Large Gas Carriers (VLGC), is 294 ships currently in operation, with approximately a further 15 under construction which equates to almost 5% of the existing fleet.

As of January 2020, the world fleet of LNGCs currently in operation numbers 536, with a further 125 under construction for delivery up to 2024, which equates to over 23% of the existing fleet. Hence, the LNG sector is very much a growth market with 58 ships delivered in 2018 and 43 in 2019. The majority of LNGCs in service or under construction are typically between 125,000 metres³ and approximately 180,000 metres³ in capacity; termed 'conventional' LNGCs.

For those LNGCs utilising a Self-supporting Spherical IMO Type B, Moss, containment system, storage capacities range from approximately 123,000 metres³ up to 182,000 metres³. For those 'conventional' LNGCs utilising a Membrane system these have storage capacities range from approximately 122,000 metres³. The commonest containment systems are GTT's Mark III and NO96. In addition, there are 15 172,000 metres³ twin-screw diesel-electric ice-capable LNGCs currently under construction at DSME in Korea. At present, there are only two LNGCs utilising a Self-supporting Prismatic IMO Type B (SPB) containment system. However, there are five 165,000 metres³ SPB single-screw 19.5 knot tri-fuel diesel-electric steam LNGCs under construction at present at Japan Marine United.

Figure 2: 'Conventional' 175, 000 metres3 4 tank Membrane containment system TS DD DFD LNGC

Currently the largest LNGCs in the world are the 14 strong Korean built Q-Max Class, where the 'Q' stands for Qatar and these are the 'maximum' size ships that can be accommodated by Qatar's LNG Terminals. They are configured with five LNG cargo tanks utilising a GTT membrane containment system giving a capacity of 266,000 metres³. These ships were completed between 2007 and 2010 and are capable of a speed of approximately 19.5 knots from twin screws powered by slow-speed direct-drive diesel-only engines rated at 21.77 MW each. Due to the fitting of liquid fuel only machinery they are not able to burn the Boil-off Gas (BoG), as is the case with most other LNGCs, and hence they are equipped with an on-board re-liquefaction system to liquefy the BoG and return it as LNG to the cargo tanks. The 31 Q-Flex LNGCs, of typically 210,000 metres³ to 216,000 metres³ capacity, are similarly fitted with liquid fuel only machinery and a re-liquefaction system.

2. LNG SUPPLY AND UTILISATION CHAIN

In 1964, the United Kingdom and France made the first large scale commercial LNG trade, buying gas from Algeria and introduced a new era of energy trading. Moving on three decades, in 2006 Qatar became the world's biggest exporter of LNG and as of 2019 only 19 countries export LNG with Qatar being the source of 25% of the world's LNG exports, Cameroon joined this exclusive club in 2018.

Originally, due to the large capital investment in land-based liquefaction and storage terminals, the specialised LNGC tonnage, and land-based reception storage and regasification facilities, the LNG trade was exclusively carried by liner services with ships on long-term charters. However, over the past decade or so a spot trade market has emerged, and with day-rates up to US$195,000 it can be a very lucrative sector.

At present the natural gas market is only about 60% of the crude oil market (measured on a heat equivalent basis), of which LNG forms a small but rapidly growing segment. Much of this growth has been driven by the need for clean fuel and some substitution effect due to the high price of oil, primarily in the heating and electricity generation sectors. To illustrate this growth, in 1970 global LNG trade was 3 million tonnes per annum (MTPA), over five decades later in February 2019 it had increased over one-hundred-fold to 393 MTPA. The 2018 trade in LNG, increased by 9.8% (28.2 MTPA) from the 2017 level, and this was the fifth consecutive year of incremental growth. Trade is expected to continue to grow substantially in the future with an increase at 6.7 percent per year in 2020 predicted.

With exports of 78.7 MT, Qatar has traditionally been the largest LNG exporter, a position it has held for over a decade. Qatar's global market share has however continued to fall to 25%, as its production remains mostly stable while exports from other markets have grown. There has been a slight shake-up in the rankings of LNG exporters, with the US jumping to fourth (21.1 MT) in 2018. Australia and Malaysia have traditionally been second and third, respectively. However, Australia will overtake Qatar, when the final 2019 figures are released; output is expected to increase to 77.5 MT compared to Qatar's expected 75 MT. Australia's output has increased significantly during 2019 due to new production coming online from *Prelude* FLNG and Ichthys LNG. Nigeria clung to the fifth position with

20.5 MT, but Russia is likely to surpass Nigeria, and possibly even Malaysia, during 2019/2020 as production from Yamal LNG increases.

As mentioned, world liquefaction capacity is currently 393 MTPA but with proposed schemes, including utilisation of Shale Gas, this could potentially be increased by up to 845 MTPA. There are currently 101.3 MTPA of plant investment sanctioned, the majority in the USA. A further growth of 24% in capacity is expected to come online in the period 2020 to 2024. [8]

As a guide, considering 'conventional' LNGCs, then 1 MTPA can on average be carried by 1.5 ships, hence an increase in production of LNG of 100 MTPA will require approximately 60 new LNGCs. With regard to this, it is worth noting that LNG output from US Shale Gas sources grew from 4 MTPA in 2012 to 34 MTPA in 2017 - almost a ten-fold increase in five years. In 2018 this output increased by a further 11%.

2.1 FLOATING LNG

There are two different applications of floating LNG: LNG Floating, Production, Storage and Offloading units (LNG FPSO termed FLNGs); and Floating, Storage and Regasification Units (FSRU). FLNG facilities are a relatively new concept, with very few in operation today, but it is realising its potential, with different technology solutions for different developments. FSRU concepts have been deployed regularly and successfully around the world over the last decade or so - advantages can include speed and affordability / scalability when local demand is small or new, and development of an onshore terminal is challenging.

2.1 (a) FLNGs

There are few FLNG applications in the world to date. In fact compared to land based plants, these concepts are small scale applications in any case, with production capacities ranging from 0.5 to 3.6 MTPA. FLNGs have been discussed for decades, since the first concept of a barge was developed in the 1950s. Only four units, the Malaysia *PFLNG Satu*, Shell's *Prelude*, Exmar's small *Tango FLNG* and the conversion *Golar Hilli*, are in operation. There are another five units either under construction or under conversion.

Offshore FLNG facilities may be exposed to harsh met ocean conditions. The concepts are usually equipped with weather-vaning anchoring and dynamic position systems. Such types of FLNG will also require LNG offloading based on Ship to Ship (STS) with units moored alongside or in tandem configurations.

FLNG's operated in more protected areas, such as near to shore or near ports, are used to produce LNG from onshore gas and eventually maybe offshore gas, supplied by a short pipeline. The locations have relatively benign water conditions since they are usually not exposed to open seas. Because they are located near shore, transfer of personnel and equipment is easier, and the accommodation and service facilities on board can therefore be limited, reducing the total Capital Expenditure (CAPEX) of the installation. In addition, the mooring equipment may be similar to a permanent mooring system for a floating installation or ship onto a jetty.

Figure 3: Large 270,000 metres^3 LNG / LPG Membrane containment system FLNG with external turret and aft accommodation for benign locations with a 'conventional' Membrane containment system TS DD DFD LNGC approaching to load cargo side-by-side

2.1 (b) FSRUs

After more than a decade of operations, FRSU solutions are considered a proven and reliable technology. FSRUs are also flexible since relocation after a period of operation in a single location is highly feasible. FSRUs have been seen as an advantageous alternative to onshore terminals, with the main benefits being the reduced cost and easier implementation. For instance, regulatory approvals may be less time consuming due to the lesser environmental impact.

The first FSRU was a new built ship of 138,000 metres^3 constructed by DSME, in South Korea, for Excelerate, and designed to offload gas on open sea conditions. This unit, delivered in 2005, was followed by other similar new builds which were equipped with the same type of regasification and mooring system, based on an internal turret and offloading buoy for the gas, which is connected to a subsea pipeline. This concept is still used in a few locations and is called the 'Gateway' concept. Another feature of this concept was that the units were weather-vaning moored to the buoy only during the offloading operations, typically less than ten days, and then disconnected and returned to the export LNG terminal to take another cargo. There are currently in excess of 20 FSRU's in operation worldwide, around 15 on order and up to 40 planned in the immediate future.

The total FSRU capacity, in terms of regasification, is relevant when compared to global regasification capacity as it represents approximately 15% of the Global LNG regasification capacity. However, the utilisation rate is lower, since for instance one third of the total modern fleet is actually operated as LNGCs (approximately 7 units). Furthermore, it must be noted that to deploy these LNGCs as import terminals, infrastructure construction would be needed onshore, such as pipeline, jetty, etc.

Figure 4: Large yoke moored 350,000 metres3 LNG Membrane containment system FSRU purpose-built design with a 'conventional' Membrane containment system TS DFDE LNGC alongside discharging cargo

3. LNG AS A FUEL

3.1 REGULATORY DRIVERS

The three main regulatory restrictions on air pollution for ships are as follows:

- EU (European Union) MRV (Monitoring, Reporting, Validation): Designed to reduce CO2 emissions from ships above 5,000 tonnes within European Waters. On 31st August 2017 a CO2 Monitoring Plan had to be submitted, 1st January to 31st December 2018 was the first reporting period, being verified on 30th April 2019 and made public in June 2019;
- IMO (International Maritime Organization) DSC (Data Collection System): Global requirement from IMO that by sometime after 2025 that there is a 30% CO2 reduction for new ships, through the introduction of the EEDI (Energy Efficiency Design Index) and SEEMP (Ship Energy Efficiency Management Plan) for measurement. EEDI was the first major step in improving the design of the ship and equipment (engines etc.) and requires a minimum energy efficiency level per capacity mile., from a 2013 reference the level was been tightened from 1st January 2015, then again from 1st January 2020 and finally from 1st January 2025;
- IMO MARPOL Annex VI: 0.5% global sulphur cap that came in to force on 1st January 2020.

3.2 OTHER DRIVERS

Clean port operations is another major driver and which is becoming much more evident as time moves forward, as is illustrated by the introduction of Cold Ironing in certain ports. To achieve this is going to require a global renewal programme with shipyards adopting cleaner fuels for their ships.

The carbon footprint of all supply chains is becoming a significant issue and will only grow as a major differentiator as the consuming public focus increasingly on this in the purchasing of everything from daily staples such as a banana or jar of coffee up to major purchases such as cars etc. This is being manifested in shipowners ordering ever larger ships, as illustrated by the fact that 22,000 TEU Container Ships are now becoming the norm and there are proposals of doubling this size in the future, as the more containers a ship carries the more the carbon footprint of each container falls.

Regulation and, as has been all too evident over the past year, public pressure will also drive change to cleaner fuels and eventually zero carbon fuels. As illustrated above, the reduction in carbon footprint will become a business imperative for the whole of the shipping industry.

3.3 LNG AS A MARINE FUEL

There are a number of alternative fuels available and proposed, but undoubtedly LNG is a, if not the, front runner as a 'transitory fuel' during the shipping industries inevitable journey to zero carbon fuels later this centuary. It is noted that even though LNG is a single carbon fuel and has a 25% saving over liquid fuels, its most significant advantage is its global availability - which will only increase in the near future.

The adoption of LNG as a primary marine fuel for conventional shipping has a number of considerations:

- Influence on ship design;
- Simultaneous operations;
- Interface of the International Maritime Organizations (IMO) International Code of the Construction and Equipment of Ships Carrying Liquefied Gases in Bulk (IGC) and International Code of Safety for Ship Using Gases or Other Low-flashpoint Fuels (IGF) Codes;
- Training and procedures;
- Risk assessment, mitigation and uniform standards.

Suppliers have been slow to define and publish the cost of LNG as a marine fuel. Global fuel pricing will emerge as supply volume increases, and it is probable that this will have geographic differences similar to the current liquid fuels market.

3.4 LNG BUNKERING INFRASTRUCTURE

However, a new LNG bunkering infrastructure is required as the existing LNG terminals are for the import and then storage of LNG which is ultimately vaporised for introduction into natural gas national grids - for which temperature is not a major issue. On the contrary, the transhipment of LNG for use as a fuel is highly temperature dependant as LNG at -157°C is not acceptable if it is to be bunkered to a receiving ship fitted with Membrane tanks. It is noted that some LNG terminals are fitted with flash tanks to guarantee a lower temperature, but maybe only down to -159°C which is still not acceptable.

3.5 ADOPTION OF LNG AS A MARINE FUEL

One of the major issues to be overcome during a retrofit is the incorporation of the new LNG Fuel Tanks on an existing ship, as there will be limited suitable locations available to the designer. Obviously, ships with large, clear open Main decks such as Oil Tankers, Bulk Carriers etc. can have LNG Fuel Tanks located on or above the Main deck. However, this is not an option on ships such as Container Ships unless significant Container Slots are lost either above and / or below the Main deck.

From above, retrofits on many ship types can, by definition, often be difficult and problematic and therefore expensive and hence uneconomic. However, there have been significant examples of existing tonnage being converted to operate on LNG as an alternative to liquid fuel, but moving forward it is likely that the numbers of ships converted will be relatively small and limited to some high-value and dedicated tonnage with which, through their design, are relatively easy to convert. It should also be noted that some recent larger ships have been designed as 'LNG ready' and delivered from the shipyard as 'fitted for but not with' the cryogenic LNG storage and systems.

However, undoubtedly the most likely approach to be adopted will not be retrofits but the introduction of new purpose-built LNG fuelled tonnage.

4. CURRENT AND FUTURE DEVELOPMENT OF LNG BUNKERING

The LNG bunkering market developed early in 2002 regionally in Norway, to deliver this new bunker fuel to small ships, such as Offshore Supply Vessels (OSVs), Fishing Vessels and coastal Ferries. However, the environmental regulations put in place by international and local regulatory bodies are now driving shipowners to build new ships or convert the existing ones to LNG fuel, displacing other traditional bunker fuels such as Heavy Fuel Oil (HFO) and Marine Diesel Oil (MDO). This choice of LNG as a marine fuel has necessitated the growing need for LBVs to support large deep-sea ships that cannot receive sufficient bunkers by road tanker.

There were 32 small scale LNGCs built in European, Japanese and Chinese yards and these have been entered into service since the early 1990's. These have capacities ranging from 1,000 to 20,000 metres^3, but none have been specifically designed and built for STS LNG bunkering operations.

There were seven dedicated LBVs and LNG Barges in operation by early 2019, and this number is expected to increase to 20 by late 2020. Initially these dedicated LBVs are being introduced to service medium to long-term charter parties that will be supplemented by ad-hoc opportunities. One LBV is designed specifically to service only one dedicated ferry service. The 12 LBVs currently under construction are between 200 metres^3 and 18,600 metres^3.

The evolution of the LNG fuelled fleet has been slow, but a number of large ships are expected to be delivered from 2020 onwards. At the beginning of 2019 a total number of 140 ships using LNG as fuel were in service, with over 160 ships on order, including at least 35 large tonnage ships. These include ultra large Container Ships, Bulk Carriers and Aframax Oil Tankers, etc. Of the 116 Cruise Ships that have been ordered in recent years, for delivery in the period 2020 to 2028, 30 of them feature LNG propulsion. This requirement has been driven by the strict local and national environmental regulations at ports in sensitive areas and by the cruise operators wishing to exhibit a greener outlook. These ships will require a larger bunkering capability in ports away from general trading routes, such as the Caribbean. There has therefore been a clear evolution since the first LNG powered ships incorporating a few hundred metres^3 storage capacity to the recent ultra large Container Ships ordered by CMA-CGM with a total LNG fuel capacity of 18,600 metres^3.

Figure 5: 7,500 metres^3 2 'cylindrical' Type C tank ballast-free TS DFDE DP II LBV *Kairos* with forward accommodation and midship and stern offloading designed to bunker a wide variety and size range of LNG fuelled ships, bunkering the *MS Visborg* in the Baltic port of Visby

The development of LNG bunkering for larger ships requires a dedicated service within a major port or adjacent ports in close proximity. Services are typically being driven by the needs of a major LNG user or producer. New bunkering hubs are also developing which will leverage existing bulk LNG infrastructure, and the majority of the world's top ten bunkering ports are now either offering LNG bunkering or have firm plans to do so by the end of 2020. Currently LNG bunkers can be obtained at 11 European ports with a further 14 ports with immediate plans to develop a service.

As the number of LNG consumers increases the number of LBVs required will need to increase to meet the growing demand. It is likely that a range of LBV types will be developed to suit individual port and regional requirements. Smaller ports may well utilise a multi-fuel capable Bunker Vessel that can carry LNG, LPG and conventional MDO / HFO in order to remain competitive.

5. LNG BUNKER VESSEL SCOPING AND DEVELOPMENT

The development and design of LBVs is a new and, for a number of reasons at the moment, a 'strange' issue! Due to the infancy of the concept, at present there no fixed and industry accepted designs because everyone requires, or perceives they require, something different. The shipping industry is currently to some degree 'feeling it's way' and 'learning by mistakes' as when many of the current LBVs were being specified, procured and built, no-one knew what the future was going to be and in some instances did not even know with any certainty how the receiving ships, to which they would be bunkering LNG, could possibly be configured.

5.1 BUNKER VESSEL OWNERS

Now, as discussed above, because of the inherent immaturity of LNG bunkering operations it is obviously a difficulty for potential shipowners to fully understand the operations that their LBV may be tasked to undertake even a in few years' time not to mention a decade or so in the future.

In addition, the technical departments of many shipowners no longer exist to the degree they did decades ago, and there are a large variety of receiving ships and operators whose requirements may not be fully known or understood.

Therefore, at this juncture it is obviously very difficult, if not impossible, for shipowners to produce a shipbuilding specification for an LBV which is as robust and mature as that they would issue to shipyards for more established ship types, such as coastal / small-scale Feeder LNGCs of a similar capacity. The only current exception to this, and hence avenue for reducing uncertainty, is if a LBV is being procured to solely bunker only one or two specific ship types the requirements for which are well known and fixed - however, such dedicated operational scenarios are the minority and if the utilisation of the asset, namely the LBV, is to be maximised it is likely it will still have to service other ships with configurations and requirements which are probably conflicting and are unknown or cannot be predicted with any certainty when the shipbuilding specification is being produced.

5.2 SHIPBUILDERS

This inability of ship owners to produce a shipbuilding specification with any certainty is a major issue for many if not most shipyards, as one of their primary goals is to propose a low risk compliant technical response which has commercial advantage over its competitors - this obviously requires a unambiguous specification.

The reason behind this is that, in many cases shipyards no longer incorporate a 'true' concept Design Office within their organisational structure and hence do not have naval architects etc. to call upon who are fully grounded and experienced in this type of 'scoping' work to develop very novel ship designs. Rather, they typically have a simplified Design Office function that purely adapts and gradually evolves existing engineered and built designs, or possibly maybe just an Engineering / Drawing Office function to support tender exercises as well as production.

In addition, due to the requirement to maintain commercial advantage in what is without doubt a 'cut-throat' industry with inherently low margins and fluctuating demand, many shipyards endeavour to focus on one or a low number of ship types and sizes in order to optimise the overall production process and hence minimise cost. This not only negates the need for a concept Design Office but also often results in designs offered by shipyards to shipowners, in response to their shipbuilding specification, being what the shipyard wants to build (to suit their production processes and throughput) rather than what the shipowner wants and needs. Therefore, shipyards do not, and often do not see the need to, fully appreciate the operational profile and novel aspects of a specific ship type.

5.3 CURRENT RELATIONSHIP

Therefore, regarding the realisation of a holistic near-optimal ship design the above can be viewed as 'the perfect storm'!

On one hand we have: the shipowner who cannot robustly bound the operations of the ship to be procured and may not have the required resources to guide design development. On the other hand we have the shipyard who may not have the resources or time to intelligently interrogate and question a shipbuilding specification, assess its validity to be transposed in to a cohesive design, and then have the ability to distil out and develop a ranked set of dominant design drivers and requirements from which a 'solution cloud' can be produced and analysed in order to robustly down-select solutions for incorporation in to the building blocks for an near-optimal operationally effective ship design.

Hence, this can be allegorised as a scenario where the 'exam question' cannot be written and 'exam candidate' has not been studying the appropriate syllabus!

5.4 REQUIRED SOLUTION

The solution to the above is not straightforward but is achievable given time, as the LNG bunkering sector matures, and with the determination of all stakeholders (trading ship operators, LBV shipowners, ship designers, shipbuilders, industry organisations, regulators, etc.) to work together in a collaborative manner.

6. LNG BUNKER VESSEL DESIGN FOR OPERATIONS

Compatibility of the LBV with the receiving ship is the key issue in designing a near-optimal LBV and it's all about alignment, so the driving force is knowing 'who' you are going to supply. Hence, a through and robust compatibility study is now the key driver in the ship design process - although typically shipyards will not like this as, as discussed above, they want to build a hull that suits their production processes. This is understandable as obviously they do not operate LBVs, or any other type ships, and hence don't appreciate, at present, LBVs very specific operational requirements! However, shipyards need to learn and they, together with shipowners, are learning quickly that LBVs are different to both traditional Oil Bunker Barges and also coastal trading / small-scale Feeder LNGCs. Hence, both parties need to understand with certainty and to some depth what the design has to achieve in operation and then undertake additional investigations, design and engineering, compared say to a small-scale Feeder LNGC, to address this.

Key three aspects driving the design of LBVs, in addition to standard ship design drivers, are:

- Parallel Middle Body (PMB) length;
- Manifold(s) location and alignment;
- Access from the LBV to the receiving ship.

As eluded to above, the design of an LBV is not simply an exercise of taking an existing design for a small-scale Feeder LNGC, adding some cranes and fenders etc. and adapting the cryogenic cargo system to export LNG. It is so much more as it is a little known very advanced ship type which is operationally and technically unique. Hence, it has with very specific and different design drivers in addition to those in standard ones associated with small-scale Feeder LNGCs.

The design and capacity of LBVs varies vastly at present due to current perceived demands, charters etc. and is obviously to some degree influenced by the current and proposed shore infrastructure. Hence, the design of LBVs will also change as local infrastructure develops as well as the design of receiving ships, to be bunkered with LNG, evolves.

Due to the complexity of LBVs, they attract a cost penalty as compare to a standard small-scale Feeder LNGC of the same capacity. Hence, in developing a robust solution for bunker operations the utilisation of purpose-built LBVs has to be closely considered and managed with care to justify the investment needed.

As a consequence, this requires some new thinking and collaboration on behalf of both the LBV owners (and the operators of potential receiving ships) and designers of an LBV. In addition, the development of an operationally robust LBV requires the, sometimes significant, adaption of current marine technologies and also incorporation and marinisation of some new ones to both the hull and its systems.

6.1 PARALLEL MIDDLE BODY

Most of the shipowners want, as is the traditional approach with Oil Bunker Barges, the shortest hull possible. The reasoning behand this is due to the requirement to potentially manoeuvre in 'tight areas' such as harbours and because of Port Dues, so as with traditional Oil Bunker Barge designs, it is often perceived that L_{OA} should be short and ideally not above 100 metres.

However, this is a contradictory and incorrect requirement as the driver that needs to be closely and accurately considered is what types and configurations of receiving ship the LBV is bunkering to and hence mooring against. For instance, consider that an LBV has to bunker and hence moor against a very large Container Ship, such as the 22,000 TEU ships currently under construction. These ships feature a forward Accommodation Block and the LNG Bunker Tanks are located below the accommodation almost on the forward tangent (Flat of Side, *FoS*) line and the Bunkering Station is further forward of this. Hence, the issue is how ensure enough parallel body, hence *PMB*, on a short LBV to safely moor alongside such a receiving ship.

Therefore, regarding principal dimensions for good ship design the dominant criteria is not so much simply minimising the length and other dimensions of the LBV, as is typical with traditional Oil Bunker Barges, but providing the required parallel body length, hence *PMB*, to facilitate safe and efficient mooring operations. The outcome of this is that quite often a longer LBV, with non-typical dimensional ratios compared to traditional coastal trading / small-scale Feeder LNGCs, will have a significantly larger operational envelope and hence potential utilisation and earning potential advantage over a slightly shorter LBV of typical dimensional ratios which may have a marginal advantage on first (construction) cost, CAPEX.

6.2 PRINCIPAL PARTICULARS

Noting the above, the LBV hull must of course be of sufficient size and configuration for:

- LNG cargo capacity and Containment System, and possibly Fuel Oil capacity;
- Cargo Configuration - number and distribution of tankage;
- Water Ballast capacity - is constant draught required / desired;
- Bunkering and Transit Environments;
- Mooring and enlarged multiple Fendering systems, and operations;
- Bunkering systems / allowance for station (manifolds) offsets, together with bunkering and crew transfer operations;
- Machinery, and Propulsion and Manoeuvring systems;
- Accommodation Block probably including enhanced crew levels over those for small-scale Feeder LNGCs; storing and associated Laydowns etc.; Life Saving Appliances (LSAs) etc.

As with standard small-scale Feeder LNGCs, and ships in general, the selection of the principal particulars is an iterative process and the designer needs to be mindful of the limiting relationships between Length between Perpendiculars (L_{PP}), B, Depth (D), Design Draught (T_D) and Scantling Draught (T_S) with regard to global strength, intact stability, roll motions, freeboard (regulatory and for survivability - both static and dynamic) etc. However, it must be noted that there may possibly be navigational constraints on L_{OA}, B and T_D, and maybe also Air Draught due to bridges etc. Hence, it is unlikely that the standard ratios typically associated with either Oil Bunker Barges or standard small-scale Feeder LNGCs will not be appropriate.

As an illustration that ship design is an often-conflicting multiple criteria problem, typically it is not desirable to have LBVs which are too 'stiff', i.e. with a too large a Transverse Metacentric Height (GM_T). This is because this may result in unacceptable motion characteristics, specifically roll acceleration. This situation would be detrimental with regard to crew and bunkering operability, and overall ship safety. Conversely, too low a GM_T is not just detrimental regarding stability but can obviously also be regarding roll angle in a seaway - it should be noted that whilst passive (e.g. bilge keels) and active (e.g. free-surface or forced roll damping tanks) solutions are available to mitigate roll, that the selection of robust principal particulars, and to a lesser degree distribution of weights (hence subsequent mass inertia as well as GM_T), in order to 'de-tune' a design from the predominant wave environment is a significantly more elegant and operationally efficient solution.

6.3 INTERNAL TOPOLOGY

Subdivision is critical not only for damage stability performance, but also for operation of the LBV regarding ballasting - noting that Ballast Water Management will have to be considered, so a 'ballast free' design may have advantages. The internal layout of the LBV will directly influence, both bunkering and other standard maritime operations and hence careful consideration of this is paramount.

6.4 HULL FORM

As a consequence of the potential *PMB* requirements discussed above, the hull form of a new-build LBV will by definition potentially be a non-standard compromise in that there may be more buoyancy than is required for the, relatively light, LNG cargo. Obviously, there is requirement to propel the hull through the water and hence a full hull form with large block coefficient (C_B) should be avoided if possible and a low B/T_D is advantageous both for resistance but also potentially propulsion etc.

However, there is an additional requirement over standard small-scale Feeder LNGCs and Oil Bunker Barges for very good slow or zero speed manoeuvrability, usually incorporating Dynamic Positioning (DP). This is because as a LBV is an IMO IGC Code ship and as such manoeuvrability is required in order to negate the requirement for Pilots and Tugs required by SIGTTO (Society of International Gas Tanker and Terminal Operators). This will require additional multiple propulsors / manoeuvring devices to be incorporated which will obviously influence the form both globally and locally. In addition, a lower T_D will be advantageous in reducing blockage effect during low or slow speed manoeuvring, which could mean a higher B/T_D is required.

Shaping of the bow can possibly be similar to that utilised on contemporary small-scale Feeder LNGCs. Whereas the stern may be more akin to other highly manoeuvrable ship types rather than trading small-scale Feeder LNGCs, and hence require a long cut-up buttock-flow stern in order accommodate the propulsors etc. and this shaping of the aft end can conveniently provide a skeg for stern thrusters (should they be necessary) and provide some directional stability. If operating in restricted waters, then a form with low-wash characteristics would be advantageous. Obviously, when developing both the bow and stern lines the *FoS* should be maximised in order to satisfy the *PMB* requirements for berthing with the receiving ship.

6.5 MANIFOLD LOCATION AND ALIGNMENT

Due to uncertainty in required bunkering operations, many existing LBVs have been designed given presumed remits to be able to bunker all ships from large high freeboard containerships, with possibly the bunkering manifolds at Main deck level, down to shore-side trucks located on a quayside. In addition, initially it was envisaged that LNG bunkering manifolds etc. would only be located in the 'open air', and hence on Main deck of the receiving ship. This was because of the perceived risks associated with potential gas releases - a valid initial assumption as all LNGCs etc. adopt this approach. Hence, in addition to midship manifolds and associated equipments, a number of newbuild LBVs had high-reach cranes and long LNG hoses and associated storage reels specified to accommodate such perceived requirements.

However, the fitting of 'Gun Port' doors located in the side shell above the waterline and the location of the LNG manifolds in way of these, as is standard for bunkering of Fuel Oil etc. on high freeboard ships, has now emerged as the preferred option for LNG fuelled high freeboard ships such as Cruise and large Container Ships etc. Moving forward it is likely that only low freeboard ships such as Bulk Carriers and Oil Tankers will typically have LNG bunkering manifolds located on Main deck. Hence, it is probable that the aforementioned high-reach cranes and long LNG hoses are unlikely to be required in the future or specified for new LBV tonnage.

The majority of LNG bunkering has been undertaken with the LBV moored side-by-side with the receiving ship. Hence, LBVs are typically fitted with midship manifolds, potentially more than one set at different heights above the waterline (or Main deck), with associated craneage to facilitate hose handling.

However, the shipping industry is now seeing the emergence of LNG bunker manifolds on the bow of a receiving ship, such as some new Feeder Container Ships, or on the very stern of a receiving ship, like a number of Ferries or some Container Ships which have Main deck mounted LNG fuel tanks located aft of the Accommodation Block. Hence, this therefore requires LBVs to have additional bow or / and stern manifolds with associated craneage etc. to facilitate hose handling - and it must be noted that these must be at the extreme bow / stern not just forward / aft of midships. Therefore, unless both the LBV and receiving ship's manifolds are aligned, namely both on the bow or both on the stern then the LBV will have to moor in a 'reverse' configuration i.e. with both ships facing in different directions which could have issues due to locations of the relative *PMB*s of the LBV and receiving ship and the location of the fenders on the LBV.

The obvious solution and logical approach to this, which would by default would eliminate the required parallel body length issue for safe mooring, is to utilise tandem mooring and bunkering. It should be noted that such solutions have been engineered and favourably viewed by LNG industry organisations. However, at present this arrangement has not been adopted even when there are very difficult and complex bunkering operations, i.e. when in open waters or when anchored and 'swinging'. In such cases very complex and costly side-by-side mooring and bunkering arrangements have currently been chosen. However, given time LBV designs may, and probably will, evolve to facilitate tandem bunkering operations.

Figure 6: 7,500 metres^3 2 'bi-lobe' Type C tank constant-draught TS DFDE DP II LBV with aft accommodation and midship and bow offloading designed to bunker a wide variety and size range of LNG fuelled ships

6.6 ACCESS FROM THE LNG BUNKER VESSEL TO THE RECEIVING SHIP

Access from the LBV to the receiving ship is obviously a major safety issue in preparing for and concluding mooring and bunkering operations. Typical solutions involve the utilisation of ladders or man-riding cranes. The current adoption of 'Gun Port' doors located in the side shell of the receiving ship is assisting this situation.

6.7 SAFETY ZONES

It must be noted that the LBVs and LNG propelled ships are designed totally independently and to different codes, namely the LNG propelled receiving ship being built in accordance with the IMO IGF Code and the LBV to the IMO IGC Code. However, once the two ships are moored together the whole dynamic changes and you have a temporary safety zone centred around the manifolds of both ships, which only exists during bunkering and changes depending on the size and configuration of each receiving ship. Unfortunately, this transient safety zone is not covered by either the ICG or IGF Codes! Rather the extent of this safety zone is governed by an analysis undertaken by the Society of Gas as a Marine Fuel (SGMF) considering both ship designs, transfer characteristics and environmental conditions etc. Everything within this safety zone is under the control of Person in Overall Advisory Control (POAC) who can be thought of as an enhanced gas engineer as the responsibility is for the transfer of LNG and hence thermodynamics must be of prime consideration.

6.8 SUMMARY

So, as highlighted and discussed above, the biggest problems for LBVs are: parallel body for safe mooring; location and number of manifolds to allow bunkering and facilitate safe operations; and safe transfer of personnel from the LBV. However, there are all the normal ship design drivers and also some others. These design attributes must be robustly established with regard to all anticipated receiving ships.

7. CONCLUSIONS

Regarding choice and utilisation of marine fuels, both regulatory and commercial priorities are working together. However, other factors now beginning to influence this, these being compliance and business advantage - as if the charterer has a choice between a perfectly capable and efficient ship and an equally efficient but cleaner ship then the latter will be chartered.

LNG is arguably the current 'transitory' marine fuel of choice in the shipping industries journey to zero carbon fuels later this century. Its implementation is not straightforward and there are a number of issues that need to be considered and overcome. However, the LNG industry is over six decades old and hence the technology is mature with associated low risk, and any issues can be engineered to produce a safe design solution.

With the adoption of LNG as a marine fuel then investment in both LBV tonnage and shore infrastructure is required. This is directly analogous to that faced by the shipping industry a century ago when liquid hydrocarbon fuel, namely oil, was introduced as an alternative to solid hydrocarbon fuel, namely coal. The incumbent issues were tackled and successfully addressed, and this change in fuel resulted in major improvements in ship efficiency. The shipping industry shall overcome the hurdles associated with the introduction of a cryogenic hydrocarbon fuel, namely LNG, and no doubt again make major associated gains in efficiency due to advances in ship design. However, on this occasion we must not just focus, as unfortunately the shipping and maritime industry still are, on the simple operational transport efficiency and emissions of a ship in isolation. The industry must assess and minimise through design the overall energy efficiency of all assets (ships) and their associated infrastructure throughout the supply chain from cradle (raw materials) to grave (scrapping and recycling) in order to consider and assess true environmental impact and sustainability in a holistic manner.

An LBV is not just a small-scale Feeder LNGC with some offloading hoses. Rather it is technically a unique and very advanced ship type with very disparate design drivers. As a consequence, this requires some new thinking and collaboration on behalf of the receiving ship operators, the LBV owners and the ship designers together with the adaption of current marine technologies and also the incorporation and marinisation of some new ones to a ship and its systems.

8. DISCLAIMER

The views expressed in this paper are those of the authors and do not necessarily represent those of the organisations with which they are affiliated and the professional institutions of which they are members.

9. REFERENCES

1. HUTCHINSON, K.W. and DOBSON, D.C. 'Floating LNG - Design Drivers and Technical Challenges', International Conference on Design and Construction of LNG Ships, The Royal Institution of Naval Architects, London, United Kingdom, 26th and 27th October 2016. ISBN: 978-1-909024-57-1

2. HUTCHINSON, K.W. and DOBSON, D.C. 'Small-scale Transhipment, Distribution and Utilisation of LNG', International Conference on Design, Construction and Operation of LNG / LPG Vessels, The Royal Institution of Naval Architects, Alba House, Clydebank, Glasgow, United Kingdom, 29th and 30th November 2017. ISBN: 978-1-909024-99-1

3. HUTCHINSON, K.W. and DOBSON, D.C. 'Near-shore floating liquefied gas storage and processing', International Conference on LNG / LPG Ships 2018, The Royal Institution of Naval Architects in association with the Hellenic Institute of Marine Technology, The Eugenides Foundation, Athens, Greece, 5th December 2018. ISBN: 978-1-909024-86-1

4. HUTCHINSON, K.W. and DOBSON, D.C. 'Small scale maritime distribution and utilisation of LNG', The Naval Architect, The Royal Institution of Naval Architects, London, United Kingdom, October 2018, Pages 42 to 45. ISSN: 0306-0209

5. HUTCHINSON, K.W. and FEATHERSTONE, R.A. 'Small scale FSRUs - a flexible solution for the future', Offshore Marine Technology, The Royal Institution of Naval Architects, London, United Kingdom, 2nd Quarter 2017, Pages 24 to 26. ISSN: 1757-7977

6. HUTCHINSON, K.W. 'Offshore FLNG - now a reality in exploiting stranded gas reserves', Offshore Marine Technology, The Royal Institution of Naval Architects, London, United Kingdom, 3rd Quarter 2017, Pages 23 to 26. ISSN: 1757-7977

7. MACKIE, G.C., HUTCHINSON, K.W. and WANLESS, D., 'Exploiting Stranded Gas Reserves: Options and Solution Development', Deep and Ultra Deep Water Offshore Technology, International Conference, The Institute of Marine Engineers and the University of Newcastle upon Tyne in association with the Society for Underwater Technology, Newcastle upon Tyne, United Kingdom, 25th and 26th March, 1999.

8. IGU World LNG report - 2019 Edition, International Gas Union, Barcelona, Spain, 2019.

10. AUTHORS BIOGRAPHY

Keith Hutchinson is a Senior Consultant for Professional Technical and Engineering Services with Safinah Group, based at their Tyneside offices in Gateshead, United Kingdom, where he undertakes commercial and offshore marine consultancy for a range of clients globally. During his over thirty year career as a ship designer and naval architect he has been constantly at the forefront of the technical development of robust but flexible concept / basic designs for a multitude of extremely novel / complex / bespoke merchant, offshore and naval vessels, both new-build and conversions - ranging from sub 20 metre high speed catamarans / landing craft to FPSOs / FLNGs in excess of 500 metres for the harshest of environments, consistently striving with clients to explore unknown multiple dimensional design spaces utilising both established and emerging technologies.

Keith graduated in Naval Architecture and Shipbuilding from Sunderland Polytechnic, now the University of Sunderland in 1990 and from the University of Newcastle upon Tyne in 1992 with a Bachelor of Engineering (BEng(Hons)) in Marine Technology with First Class Honours in Naval Architecture. Having worked in both Shipbuilding and Shiprepair yards on the River Tyne, he joined Swan Hunter Shipbuilders in 1990 and in 1992 was appointed as a Senior Designer within the Naval Architecture Section of the Basic Design Department. Having joined newly established marine design consultants Armstrong Technology Associates (ATA) as a Naval Architect in 1995 he was soon seconded to the Newcastle University's Engineering Design Centre (EDC) for over three years as a Research Associate developing decision support tools for the evaluation, optimisation and selection of safer RO-RO ship designs and other marine vehicles. Returning in 1998 as a Naval Architect he, over the next twenty years, rose to be a Senior and then the Principal Naval Architect at Babcock's Tyneside Ship Design office, who had acquired Armstrong Technology in 2000, and finally was overall Engineering Manager responsible for the coordination and management of all multidisciplinary engineering teams formulating design solutions together with cost estimating and investment appraisals. He was also Chief Engineer for Commercial Marine within Babcock in the United Kingdom, and hence as senior Technical Authority was responsible for all governance aspects relating to whole ship design, engineering etc., in addition to being Subject Matter Expert (SME) for the Naval Architecture domain, hence Company Naval Architect, across all Babcock offices in the United Kingdom from 2011 responsible for hydrostatic, dynamic, arrangement and regulatory evaluations etc.

Keith has published over fifty technical papers and articles on naval architecture and ship design etc. in addition to being invited to give over forty presentations on aspects of ship / offshore vessel design and performance. He is a Member of Lloyd's Register's (LR) main merchant ship Technical Committee (TC), Offshore Technical Committee (OTC), Naval Ship Technical Committee (NSTC), and their Rules Human Factors (HF) Panel. He is also an industry 'Briefing Member' for the United Kingdom's Delegation to IMO's Ship Design and Construction (SDC) Sub-Committee. He is a Chartered Engineer (CEng), Chartered Marine Engineer (CMarEng), Fellow of The Royal Institution of Naval Architects (FRINA), Fellow of The Institute Marine Engineering, Science and Technology (FIMarEST) and Fellow of The Society of Naval Architects and Marine Engineers (FSNAME). He is active in the North-East Coast Joint Branch of IMarEST and RINA for over two decades and is Honorary Treasurer, having previously served as Chairman and Vice-Chairman. He has been a Professional Review Interview (PRI) Assessor for membership of RINA and IMarEST and registration with the Engineering Council (EC) in the United Kingdom for over fifteen years. He is a founding Committee Member of SNAME's recently inaugurated Western Europe Section (WES) and is Technical and Research / Papers Chair, having previously served as Membership Chair.

David Dobson is Head of Professional Technical and Engineering Services, and also Protective Coatings, with Safinah Group, based at their offices in Gateshead on Tyneside, United Kingdom.

His career began in 1973 as a Ship Draughtsman with Swan Hunter Shipbuilders on the River Tyne. He attended Newcastle College, Sunderland Polytechnic and Newcastle University gaining a Bachelor of Science (BSc(Hons)) in Naval Architecture in 1983. He quickly joined the management team in Swan Hunter taking up various positions in production, production engineering and project management. He gained a Masters of Business Administration (MBA) in Strategic Management in 1992. He left Swan Hunter in 1993 and subsequently worked in a variety of Offshore, Engineering and Shiprepair companies as General Manager and Director. He joined Babcock in 2007 and was Programme Director for the commercial marine and oil and gas business stream over an eleven year period.

David is a Chartered Engineer (CEng) and was awarded Fellow of The Royal Institution of Naval Architects (FRINA) status in 1996.

